Technology and the Labour Proce

Studies in Society

Titles include:

The Migrant Presence Martin
Race, Class and Rebellion in the South Pacific Mamak and Ali
Children and Families in Australia Burns and Goodnow
Johnsonville Pearson
Australian Community Studies and Beyond Wild
Open Cut Williams
The Ethnic Dimension Martin/Encel
The Aging Experience Russell
Australian Cinema Tulloch
Strikes in Australia Waters
Class Consciousness in Australia Chamberlain
Heathcote Wild
Practice and Belief Black/Glasner
Law, Capitalism and Democracy O'Malley
Eclipse of Equality Pearson and Thorns
The Family in the Modern World Burns, Jools and Bottomley
Medical Dominance Willis
Women, Social Welfare and the State Baldock/Cass
Ideology of Motherhood Wearing
Indians in a White Australia de Lepervanche
Confronting School and Work Dwyer, Wilson and Woock
Ethnicity, Class and Gender in Australia Bottomley/de Lepervanche
The Faith of Australians Mol
Social Theory and the Australian City Kilmartin/Thorns/Burke
The First Wave Lewins/Ly
Surveys in Social Research de Vaus
Australian Ways Manderson
Custody and Control Asher
Where it Hurts Russell and Schofield
Caring for Australia's Children Brennan and O'Donnell
Shaping Futures Wilson and Wyn
Health Care Issues Bates/Linder-Pelz
A Gender Agenda Evans
Regulation and Repression Edwards
For Freedom and Dignity Metcalfe

Studies in Society

Technology and the Labour Process:

Australasian Case Studies

Edited by
Evan Willis

Sydney
ALLEN & UNWIN
Wellington London Boston

First published in 1988

Allen & Unwin Australia Pty Ltd
An Unwin Hyman Company
8 Napier Street, North Sydney, NSW 2059 Australia

Allen & Unwin New Zealand Limited
60 Cambridge Terrace, Wellington, New Zealand

Unwin Hyman Limited
15-17 Broadwick Street, London, W1V 1FP England

Allen & Unwin Inc.
8 Winchester Place, Winchester, Mass 01890 USA

National Library of Australia
Cataloguing-in-publication entry:

Technology and the labour process.

Includes index.
ISBN 0 04 337008 X.

1. Labour supply – Australia – Effect of technological innovations on – Case studies. 2. Technological innovations – Australia. 3. Work – Technological innovations. 4. Industry – Technological innovations. 5. Labour supply – New Zealand – affect of technological innovations on – case studies. 6. Technological innovations – New Zealand. I. Willis, Evan, 1952– (Series: Studies in society (Sydney, N.S.W.)).

331′.0994

Library of Congress Catalog Card Number: 88–71405

Set in 10/11 Times Roman by Graphicraft Typesetters
Printed by Kim Hup Lee Printing, Singapore

32

Contents

Contributors

Dawn Butler lives and works in Brisbane. She has recently completed a doctoral thesis at Griffith University on 'Secretarial Workers: Their changing labour process and social relations'. Her involvement in the secretarial field has spanned more than twenty years, as an educator and as a secretary. Her current activities are concerned directly with secretarial education.

Peter Cammock is a lecturer in industrial relations in the Department of Business Administration, University of Canterbury. Previous published works are in the areas of apprenticeship and religious influences on entrepreneurial behaviour. He is currently researching dimensions of managerial performance.

Jeanne Daly is the research officer on a joint St Vincent's Hospital/ La Trobe University study in the field of medical technology assessment. She is also completing a doctoral study in this field at La Trobe University. She has co-edited the book *Technologies in health care: policies and politics*. An earlier and extended version of this chapter was presented to the 5th annual conference on the Organisation and Control of the Labour Process, in Manchester in April 1987.

Richard Dunford is a senior lecturer in organisation studies at Kuring-gai College of Advanced Education. He has published in a range of journals including *Social Studies of Science, Prometheus* and the *University of New South Wales Law Journal*, and in collected readings. Current work includes a book on organisational behaviour and a study of industrial relations in small business. An earlier and extended version of this chapter was presented to the 5th annual conference on the Organisation and Control of the Labour Process, in Manchester in April 1987.

Raelene Frances is a lecturer in history at Murdoch University. She has published several articles on the history of prostitution and an article on work process in the Victorian clothing trades. This study is part of a doctoral thesis which looks at gender and the labour process in the Victorian boot, printing and clothing industries in the period 1880–1900.

Kerr Inkson is associate professor of management studies at the University of Auckland. He has published two books and over 50 papers, mainly in the fields of organisational behaviour. He is currently involved in major projects on the management of voluntary organisations and on corporate leadership in New Zealand. The attitude survey work reported in this study was supported by a grant from the University of Otago, and the 'pelt-stripping' study by a grant from the University of Auckland.

John Mathews is currently a project director in the Resources Co-ordination Division of the Victorian Ministry of Education, working on technology and industry development related issues. He was formerly head of the Working Environment Policy Division of the Victorian Department of Labour, and from 1981 to 1984 the foundation head of the ACTU–VTHC Occupational Health and Safety Unit, a national advisory body to the Australian trade unions. Much of the text of this chapter was written while the author was a visiting fellow in the department of humanities at Deakin University.

Peter McGraw is a lecturer in industrial relations at Kuring-gai College of Advanced Education, and has published in a range of journals including *Human Resource Management Australia* and the *Journal of Industrial Relations*. Current work includes a study of industrial relations in small business and consultation in the area of negotiating skills.

Len Palmer is a lecturer in communication studies at the Darling Downs Institute of Advanced Education, Toowoomba, Queensland, where he specialises in communication technology. His contribution here is drawn from his doctoral research on Telecom Australia. His current interests are in the areas of popular culture, cross-cultural relations, organisational studies and communication technology curricula.

Belinda Probert is senior lecturer in sociology at Monash University. She is author of *Beyond Orange and Green: The Political Economy of the Northern Ireland Crisis.*

Rosslyn Reed is a lecturer in social and organisational studies at Kuring-gai College of Advanced Education. She is completing a doctoral study of technological change in a Melbourne newspaper organisation. This study was supported by a doctoral research award from the Bureau of Labour Market Research. Steve Frenkel provided helpful comments.

Ern Reeders is a lecturer in social science, teaching in sociology and careers education courses at the Royal Melbourne Institute of Technology. He has been conducting research into the position of young people in school and work since 1980. This research was supported by grants from the Research Committee and the Faculty of Humanities and Social Science at RMIT. Helen Marshall provided helpful comments.

Judy Wajcman is senior lecturer in sociology at the University of New South Wales. She is author of *Women in Control: Dilemmas of a Workers Cooperative* and co-editor with Donald MacKenzie of *The Social Shaping of Technology*. The research was supported by grants from the Australian Research Grants Committee, the Bureau of Labour Market Research, and Special Research Grants from Monash University and the University of New South Wales. Warwick MacMillan was research officer on the project and Frances Stadden assisted with data analysis. Assistance is also acknowledged from Caroline Alcorso, Pauline Garde and Stuart Rosewarne.

Roger Wilkinson is a tutor in sociology at La Trobe University. He is completing a doctoral study on the diffusion of Japanese management practices into the vehicle building industry. This study was supported by grants from La Trobe University School of Social Sciences.

Evan Willis is senior lecturer in sociology at La Trobe University. He is author of *Medical Dominance* and co-editor of *Technologies in health care: policies and politics*. His current work includes projects on medical technology assessment, occupational health and safety and trade union responses to technological change.

Preface

What has come to be known as the 'labour process debate' has rejuvenated the sociological study of work in the last decade and become an exciting area of intellectual debate and research. The aim of this book is to help students understand both the original Braverman form of that debate and the variety of directions in which the concerns raised have developed. Researchers at the forefront of these developments in the Australian and New Zealand context have been asked to provide case studies demonstrating how labour process theory is being utilised to research the contemporary organisation of work, particularly the process of technological innovation.

Thanks are due to a number of people who assisted with its preparation in various ways. These include John Martin, Jim Kitay and Judy Wajcman, who made helpful suggestions, as well as Helen Cook and Barbara Matthews, who assisted with preparation of the manuscript.

1 Introduction

EVAN WILLIS

It is by now a common observation that the work we do is central to our sense of social identity—of who we are—and to human existence in helping us to feed, clothe and house ourselves. When two strangers meet, one of the first things they ask in order to locate each other's position in the social world is what sort of work they do—what they do for a living.

Yet what we mean by the notion of work is not always clear. Commonsense definitions of the concept tend to equate it with paid employment—that is, in the marketplace for money. However, this negates a huge amount of domestic labour in the form of housework and also a large amount of voluntary work. Work indeed can be any purposive social action, paid or not, involving obligation or not. How work is defined will affect how related concepts are defined. If work is not tied only to payment then occupation cannot be either. Hence 'poet' can be an occupation, however commercially unsuccessful it may be, as can 'student'. Likewise the concept of leisure becomes problematical: do the unemployed have leisure? Do housewives have leisure?

The focus of this book is not the entire range of social activities which can be called work, but with a substantial part of that phenomenon. The emphasis is placed on work as a social activity insofar as it entails an employment relationship—that is, a contractual arrangement by which a wage or salary is paid in return for a specified amount of labour applied to a task or range of tasks. For most people, this is how a 'living' is made or a livelihood gained, either directly through this employment relationship or indirectly through dependence on the employment of a spouse or parent.

Work is not a static phenomenon, however, but one which undergoes constant change. There have been several important changes in the nature of work, and the way it has been organised in Australia and New Zealand over the decades since the Second World War. These include the dilution of the male, Anglo-Saxon dominance in workforce composition, at least at the lower levels, by the entry of women and immigrants into the paid workforce; the continuance of levels of unemployment (i.e. insufficient work for those who seek it) at levels not experienced since the 1930s, despite intensive governmental efforts at job creation; and the transformation of work by technological innovation. This last change is the main focus of this book.

Much has been written about the process of technological change, a lot of it limited by a very narrow conception of what technology is. Here it will be used, as it was by Dickson (1974), to refer not only to the tools and machines used by individuals as a means of altering their physical environments, but also to the social relationships implied by their use. In other words, technology has both a material and a social component. For example, domestic video cassette technology comprises both the video cassette recorder itself and the social relationships, such as those involving copyright on recorded programs, implied by its use. Both the material and the social components of technology must be considered when examining the process of technological innovation. Hence we must examine both the social shaping of technology in the design, etc. of the tools and machines as well as the impact of its innovation.

Of course, technological innovation has affected the work people do throughout history, particularly since the 'industrial revolution' associated with the early development of capitalism. Arguably, though, the rate of change as a result of technological innovation has accelerated in recent decades, largely due to microchip technology. Scarcely a workplace involving paid or unpaid work has remained unaffected by its introduction.

These developments in the work people do (or aren't able to do) are probably the major changes in work in Australasian society over the last few decades. Of course, they are not unique to this part of the world, but are being experienced to a greater or lesser extent everywhere. Together they provide much of the raw material for study by social scientists interested in the phenomenon of work. These include sociologists, economists, historians and psychologists, amongst others in the fields of industrial sociology and psychology, labour economics and history, and industrial relations. Their approaches to studying work vary greatly, however. A distinction can be made between applying these social science insights at work and applying them in the study of work. Using sociology as an example, we can distinguish between the sociology *in* work and the sociology *of* work. The former (*in* work) has tended to have a managerially oriented focus about how work can 'best' be organised, which has usually meant for maximum productivity. It involves developing insights to use inside organisations, an approach which is shared with much of industrial psychology and industrial sociology.

The latter approach (*of* work), by contrast, has been more of the 'outsider looking in' variety, directed not so much at managerially oriented concerns but at the implications of forms of work organisation for the individuals concerned and in the wider society. It has tended to be more critical in focus and to operate with a different set of assumptions than the sociology in work approach. Whereas the sociology in work approach has largely operated with an assumption of consensus about goals and a shared interest between the various participants, the sociology of work tradition operates much more with a conflict assumption between the various participants and the lack of a commonality of interests.

One way to characterise the difference between these two approaches is to focus on the questions they ask about work. For those interested in the sociology in work approach, the key question is one of *co-ordination*: how can the various components of an organisation be co-ordinated to achieve organisational goals? For those interested in the sociology of work, by contrast, the key question is one of *control*: how can the various components of an organisation be controlled so as to permit realisation of managerially defined goals?

Most writing by social scientists in general and sociologists in particular has been of the in-work approach and only recently has the more critical of-work understanding been rejuvenated. This book is situated more within that latter tradition and the papers analyse a number of areas of the workforce in relation to what has come to be known as the labour process tradition.

The labour process

This tradition has focused more on the labour process than on work itself, because the labour process is the way in which work is organised. Central to an understanding of this organisation of work is an understanding of the system of economic relationships which provides the framework in which economic and social activity takes place. Such a materialist assumption directs attention to the fact that Australia and New Zealand are not only Western, advanced, modern, industrialised societies, but they are also capitalist societies

A detailed account of capitalism is beyond the scope of this book (see Jalée, 1977, for an introduction), but for our purposes it can be understood as a system of commodity production directed at capital accumulation. Such commodities, either goods or services, are produced in order to realise and contribute to profit. In such a system, relational class groupings of the population, defined in terms of their relationship to the means by which these commodities are produced (land, raw materials, etc.), have emerged. There is a basic inequality in the distribution of these resources in which some, the minority, either own or control (through management) these productive means. Others, the majority, have only their labour power (potential) to attempt to sell to the owners and controllers in return for a wage or salary as a livelihood. A major structural feature of capitalist societies is thus a relationship of domination and subordination between the dominant class and the working class, as the latter is subjected to the systematic pursuit of profit by the former. There is also a third, intermediary, class grouping subdivided into two sub-classes—the old middle class comprising small business persons, farmers, etc., and a new middle class of professionals who are important at the level of maintaining capitalist social order because they produce what Gramsci (1971) calls 'ideological hegemony' by acting as organic intellectuals to the dominant class. The medical profession, for example, forms part of this class grouping because of its role in writing medical certificates, conducting pre-employment screenings and so on.

These three classes are defined in terms of their position in the productive process, not like layers in a cake but in relation to one another. The nature of the capitalist economic and social system structures the employment relationship through which the labour process occurs. Another way of saying this is that work is a relationship of power for most people in our society; that it is structured in terms of a relationship of domination and subordination. As is sometimes said, 'democracy stops at the factory gate'. Such power, however, is not randomly distributed between individuals but structured into classes.

Such a tradition of study goes back to the powerful analysis of capitalism of the nineteenth century political economist, philosopher and sociologist Karl Marx (1967). Subsequent studies in this tradition have been quite numerous, but little further analysis on the labour process itself appeared until the 1970s. In the meantime, in academic circles at least, there occurred a separation of concerns with the political economy of work, which was a feature of the labour process under capitalism, into separate academic disciplines of sociology, economics, politics and psychology. With it came a loss of the critical awareness of the implications of the capitalist organisation of the labour process for individuals.

The revitalisation of this tradition came in 1974 with the publication of Harry Braverman's *Labour and Monopoly Capital: The Degradation of Work?* This study served to reorient and reintegrate the sociological study of work towards the more traditional nineteenth century concerns of political economy. Braverman's work has stimulated a great deal of research and analysis and will be judged one of the most important books published in the social sciences in the second half of the twentieth century. As Littler (1987: 2) has argued:

> Braverman's major contribution . . . was to smash through the academic barriers and offer the potential for the birth of a new and integrated approach to the study and history of work. An integrated approach which provided an (apparent) coherent theoretical framework and which also directed attention to a range of issues and problematics both within and beyond the workplace, which had previously been seen as unrelated. Braverman attempted to re-integrate a consideration of the division of labour, technology and management methods with an analysis of occupational structure, class structure and the phases of capitalism.

The publication leads to a resurgence of interest in the labour process and the questions posed by Braverman have provided a massive agenda for further investigation, though mainly in the Northern Hemisphere. In the Australasian context, at least until recently, the tradition of research in this area has been relatively limited. A whole issue of the *New Zealand Journal of Industrial Relations* (December, 1984) and the substantial theoretical treatise by Campbell (1983) constitute the main items. Only more recently has interest flowered, some of the results of which, in the Australian and New Zealand context, are presented in this collection of readings.

In the decade since Braverman's work, many of the answers he gave to the questions posed have been subjected to justified criticism. Even the most glowing reviewers of his work have argued that some of his central ideas were overstated. Considerable refinement of the conceptual tools used in the labour process approach have occurred and the focus of study has moved well beyond Braverman's original concerns. None of this diminishes the importance of his seminal work, however; indeed, the argument to be made here is to reiterate the comment by Paul Sweezy in the original introduction to Braverman's book, that the importance of his work lies not so much in the answers given but in the questions asked. (Sweezy 1974: Introduction). The papers in this collection all fall within this broad labour process tradition and represent an expression of the momentum generated by Braverman's work. Not all of the authors are sympathetic to the particular answers Braverman gave but all are, to a greater or lesser extent, working out the agenda set by Braverman in the context of historically specific Australian and New Zealand capitalist societies.

The Braverman thesis

Harry Braverman's stimulus in writing *Labour and Monopoly Capital* (1974) seems to have been a desire to buy into the debate which had developed about the implications of changes in the nature of work that people did, particularly in relation to the introduction of new technology. A view had been espoused by a number of what have come to be called 'post-industrial' theorists such as Daniel Bell (1973). They argued that the effects of technological innovation on people's work had been overwhelmingly positive, that new technology had eliminated much work that was repetitive, dirty, dangerous and boring, replacing it with clean, safe, interesting jobs so that the experience of work for most people was better as a result.

It was this view which Braverman specifically challenged. As the subtitle of his work ('The degradation of work in the Twentieth Centruy?') suggests, he argued that changes in the nature of work under capitalism in the twentieth century, of which technological innovation was an important part, far from leading to improvements in the quality of the nature of work, had degraded it. In order to sustain his argument he traced it back to the nature of (monopoly) capitalism itself. He argued that, to understand what was happening to work, or the transformation of the labour process, it was necessary to examine how capitalism itself operated and had changed. The imperative of capital accumulation in order for individual entrepreneurs to survive in a competitive economic environment led to technological innovation in the workplace. Such innovation by entrepreneurs, however, as Marx before him had argued, was not directed only at some abstract notion of 'efficiency', but also at better control over the workforce. The consequence for workers was 'deskilling', and their work was degraded as a result. In drawing this conclusion, Braverman also effectively argued for the return of the analysis of classes and class struggle to centre stage. As Ramsey (1985: 53) commented, he served to reintroduce the political economy of class to studies of the workplace when such a perspective seemed to have retreated to analyses of the State and of social inequality.

In understanding capitalism, Braverman argued (as Marx had

done before him) that the labour process takes place in the context of conflict over the generation and appropriation of surplus value. As a result, the central dynamic of the labour process is the realisation of labour power—the transformation of labour power *potential* into *actual* labour. For the working class, who have only their labour power to sell in order to make a livelihood, the task is to find an employer who will engage that labour power potential or capacity to work, be it physical strength, skill or knowledge, in return for a wage or salary. For the employer, the task is to transform the labour power potential into actual labour in order to provide wages and make a profit. Hence the role of management is to 'realise' labour power by not only organising production and the supply of raw materials, designing job tasks, etc., but also creating mechanisms (supervisors, etc.) to ensure workers perform what is often called a 'fair day's work'. So-called 'managerial prerogative', then, is to both organise the process of production and to ensure control and surveillance in the performance of the tasks. Together these features make up the capitalist labour process, which is central to understanding any capitalist society.

Braverman's answer to the question of how work capacity was translated into wages and profits, it is important to note, was *coercion.* Unlike other social science approaches, which assumed consensus over the organisation of work, Braverman assumed that recalcitrant workers had to be coerced into realising labour power potential, and that (class) conflict resulted from such a process. The analogy often used to describe such a process is to liken the contract between labour and capital to a rubber bag: the amount of work achieved depends upon the extent of coercion (Edwards, 1979).

Braverman's argument was that the labour process was designed by employers to further both these aims. The logic of capitalism was expressed by 'scientific management', a philosophy and an ideology (including material practices) which Braverman argued was developed and expressed through Taylorism; that is, by a management system originating with the American management consultant Frederick Winslow Taylor at the end of the nineteenth and early twentieth centuries. Subsequent management strategies such as Fordism (i.e. from Henry Ford) and the human relations approach were basically modifications of Taylor's original management approach whereby different methods were utilised to achieve the same end: the coercion of recalcitrant workers into realising labour power potential sufficiently to sustain capital accumulation. Management thereby mediated the relationship between the organisation of the individual enterprise and the requirements of capitalism as a whole, by devising and implementing work design in such a manner as to bring about control over the workforce. So called 'scientific' (that is neutral) management with its time-and-motion studies was demystified by Braverman as representing basically a class ideology, and not neutral at all. In so doing, Braverman rescued the sociological study of work from organisational analysis and industrial psychology and reoriented it towards economics and history.

For Braverman, Taylorism (the expression of the transformation of the capitalist labour process) had three key features. The first of these was the *dissociation of the labour process from the skills of the worker*, whereby management or its agents gathered together all the knowledge necessary to perform the production process. The aim was to make the labour process independent of craft tradition and workers' knowledge. Secondly, it involved the *separation of conception from execution*, the classic mental–manual labour split. Rather than workers planning their work tasks and then doing them, the two aspects would be divided as much as possible, so that workers would carry out a simplified range of tasks planned for them by management. Thirdly, Taylorism involved the *use of management's monopoly over knowledge to control each step of the labour process and its mode of execution*. This involved systematic pre-planning by management of all elements in the labour process, through 'work study' or 'time-and-motion studies' in which the workers would be told exactly how to complete a limited range of tasks in the production process. Workers would thus lose the responsibility for planning, creating and initiating their own work. Incorporated into this stage of job design was the *Babbage* principle, aimed at cheapening the cost of production. Named after the early British industrialist Charles Babbage, this principle reduced production costs because skilled workers could be retained in only those areas of the production process where their skills were actually needed. Parts of their former complete work tasks which were of a more routine nature could be paid for at a lower rate. By being able to plan and specify each step of the production process, management was able to pay for each task in the process exactly according to the amount of skill required.

Deskilling came about in particular through the separation of conception and execution, which reduced the amount of skill applied by the worker in the form of knowledge, responsibility and judgement. As is often said to the workers. 'You're not paid to think, you're paid to work.'

The effect of Taylorism was to divide up the labour process as much as possible, thereby increasing the division of labour. There were two related reasons for this: firstly, production could be increased by specialisation, as Adam Smith (1976: 110–11) analysed in his famous example of pin manufacture. Equally important, however, was the potential for control which existed in a highly specialised workforce—an application of the 'divide and rule' principle. Such a division of labour which Braverman held to be a result of the capital accumulation process, had the consequences for individuals of deskilling and the degradation of work. The classic example of this form of degraded, alienating work in modern society is probably car assembly, the endless performance of a tiny proportion of the work of assembling motor vehicles (see Beynon, 1973). Another cogent example comes from the study of BHP's Whyalla shipyard in South Australia by Kreigler, who argues of the tradesmen employed there that:

> [many] were shipbuilding specialists (who) maintained that they were given only the most basic and repetitive tasks and were rarely given the opportunity to perform them without excessive and unnecessary interference from supervisors. They explained they were not permitted to work from technical drawings or even to make minor decisions as to how the job should be executed. Denied all responsibility, challenge and initiative in their jobs, workmen appear to have lost interest in their work, their skills have been eroded over the years and they have lost confidence in their ability to work in industries where they may be required to apply all aspects of their trade (Kreigler, 1979: 269–70)

For Braverman, the development of the capitalist labour process was not just a product of some technological imperative, nor of some general process of modernisation. Rather, it was the product of a class society. Put another way, increasing specialisation and division of labour had to be understood in the context of class relations.

Technological innovation

Of particular interest in this book is the application of Braverman's analysis as a framework for understanding the interaction between technology and work. While scientific management provided the ideological basis for the co-ordination and control of labour, the practical basis came from technological innovation. By this process, the skill and knowledge which allowed workers to exercise some measure of control over the labour process (such as the pace of work) could be expropriated and placed in the hands of management. A central theme of Braverman's analysis was an attempt to demystify the process of technological innovation and its implications for the design of production. Braverman took direct aim at what has come to be called technological determinism, an attitude perhaps best summed up by the phrase 'we have no choice but to introduce this technology'. He argued that there is nothing inevitable or immutable about the march of technology, and that the conventional explanation for the introduction of new technology in terms of competition between employers leading to the search for greater 'efficiency' by cheapening the costs of production is only half the story. The other half is the search for better and more reliable forms of control over the workforce. According to Braverman, because the division of labour developed in the design of work was based upon the Babbage principle, then the issue of control is inseparable from the pursuit of efficiency.

Efficiency, furthermore, was not a neutral concept as the proponents of scientific management had claimed; what is most efficient for employers (such as closing down a plant or asset-stripping following a takeover) may well not be the most efficient action for the workers, who thereby lose their jobs. In other words, the process of technological innovation revolves around not only the competition between employers but also the struggle between employers and employees. Likewise, the questions to be asked about the interaction between technology and society are not only those deriving (as technological determinism would hold) from the impact of that interaction (i.e. after the event). Also important is the question of how the design of technology permits greater monitoring of and control over the workforce—that is, the social shaping of technology (see, for instance, MacKenzie and Wajcman, 1985). Technological innovation from this point of view is not an autonomous process 'impacting' upon us, but a reflection of the attempts of one class to control others with technology as the means.

The Braverman critiques

Since its 1974 publication, Braverman's work has been the focus of a great deal of critical attention by others. More than a decade later, it is now fairly clear that the importance of the book lies more in having identified the questions to be asked than in the answers provided. Problems have been identified with aspects of much of Braverman's work and it is these critiques which will be briefly examined here. Two lines of critique will be considered. One relates to the problems which have been identified with what it was assumed Braverman *didn't do*; that is, those issues, considered by others to be important to a fully adequate and comprehensive analysis of the labour process, which are not analysed or only minimally analysed. The other problems result from what he *did do*; that is, with his analysis, and in particular with the conceptual tools used to make that analysis.

The first omission of Braverman's that is considered to have diminished the applicability of his account is his purposely one-sided account of class formation. This stresses only the *objective* structural mechanisms derived directly from the process of capital accumulation which operated to reduce or destroy individuality. As a number of writers, such as Zimbalist (1979), have argued, Braverman was aware of the ability of workers to assert their subjectivity and individuality by reacting to management control, either individually or collectively, through struggle, resistance or contestation. Nonetheless, the omission of this side of the story makes his account look rather one-sided, as if management always achieves the ends it seeks (see Stark, 1980). The effect, as a number of writers have argued (e.g. Littler, 1982: 28–30), is that Braverman's analysis has functionalist strains, as if reorganisations of the labour process are the outcome of conscious design rather than the products of struggle and resistance. In many case studies, management has been found to be less than omnipotent; the history of the labour process has been found to be not the automatic advance of capital, but as Ramsay (1985) argues, contested and on occasions likely to backfire. Workers' resistance to Taylorism, in other words, cannot be relegated to a minor, derivative role but is crucial to shaping the labour process itself.

An important line of post-Braverman development to overcome this omission has been argued by Burawoy (1985), who returns some of the voluntarism and agency to the analysis of the labour process by stressing the conditions under which *consent* may emerge. The division of labour is given not only by the structures in which individuals operate or bear (e.g. supervisor, machine operator), but is also negotiated to a certain extent within these boundaries. Hence Burawoy (1985: 37–9) argues that *games* emerge in respect of technology, and between workers, as adaptations to the nature of capitalist work, and to conceal the nature of production relationships. Consent, it should be noted, is not the same as a consensus of interests between management and workers; the rules of the games have definite boundaries and a not-too-obscure aim of increasing production.

An account such as that developed by Burawoy also needs an analysis of another aspect of social order essential to capitalist societies in general, and the labour process in particular, which is largely absent from Braverman's work except in his analysis of

Taylorism. According to Burawoy (1985: 44), Braverman discounts ideology as a factor essential to the study of capitalism. Following Gramsci, Burawoy argues that the arbitrary despotism of the early phase of capitalism has been replaced by hegemonic control:

> Workers must be *persuaded* to cooperate with management. Their interests must be coordinated with those of capital. The despotic regimes of early capitalism, in which coercion prevails over consent, must be replaced with hegemonic regimes in which consent prevails (although never to the exclusion of coercion). (Burawoy, 1985: 126)

Not only is ideology an essential analytical tool undeveloped in Braverman's analysis, but so is the *State* or, more particularly, specific capitalist states, according to number of writers (e.g. Burawoy, 1985). Braverman briefly deals with the role of the State in the context of social co-ordination (1974: 269) but does not develop it. The consequence is to direct attention away from the sort of State intervention which occurs, for example, via the conciliation and arbitration system. Such intervention has been important in shaping the labour process in specific capitalist states such as Australia and New Zealand.

The other line of critique of Braverman's work has related to what *he did do* and to the perceived inadequacies of the analytical tools he developed to analyse the labour process. The first of Braverman's analytical tools which has been the subject of critical comment is the concept of *deskilling* and, with it, *skill* in general. One rather obvious line of criticism has been to argue that technological innovation in the development of the labour process may deskill workers in some areas but *enskill* others; that is, also create the demand for new or different skills. Microchip computer technology is often used as an example. While it may deskill the metalworker operating automatic computer-aided manufacturing (CAM) or numerical control (NC) machines (see Jones, 1982), it also creates demands for new skills from computer programmers, etc. This line of criticism, however, is based on a misunderstanding of Braverman's work which is not static, as such a criticism would imply, but historical and dynamic. As Zimbalist (1979: xvi) argues:

> mechanisation and deskilling of work in one economic sector implies that new processes are evolving in another (backwardly linked) sector. These new processes and techniques bring with them the demand for workers with new or initially scarce skills. It is not until a later stage of their development that these new processes and techniques become themselves subjected to rationalisation, job fragmentation and mechanisation.

In other words, 'second and third generation' technology may have similar consequences for the recently enskilled workers to those which its 'first generation' predecessors had for the shop floor.

More serious critiques of the concept of skill and deskilling argue, however, that the concepts require considerably more careful unpacking to be understood properly, and a considerable amount of analysis has shown Braverman to have a rather monolithic and mechanistic notion of deskilling (e.g. Lee, 1982; Salaman, 1982; Coombs, 1985) which contends that employers always attempt to deskill workers and workers have never managed to prevent this. Likewise, the concept of skill is a complex one; the kind of skills recognised in the workforce is highly specific historically and culturally. For one thing, skill definitions have been shown to be extensively saturated with sexual bias (see Cockburn, 1983; O'Donnell, 1984). Dexterity is a good example, highly valued in male brain surgeons but not in female clothing or food packaging workers. In an attempt to articulate the complexity of the concept in response to the perceived weaknesses of Braverman's account, Littler (1982: 7–11) has devised a three-fold classification of conceptions of skill: as an objective characteristic of work routines and job knowledge, as job autonomy, and socially constructed as social status on the basis of custom, tradition and collective organisation.

The second concept central to Braverman's analysis which has been subjected to critical comment and modification has been that of *control*. Kelly (1985: 44), for instance, has criticised the manner in which control is conceptualised in unidimensional and zero-sum terms in that adaptation as well as resistance can be strategies of response to the attempted imposition of control. Others have unpacked and reconstructed Braverman's notions. Friedman (1977a, 1977b), in his analysis of the English context, draws a distinction between *direct control* as outlined by Braverman and *responsible autonomy*, a form of limited job control (through a unity of conception and execution), in an attempt by management to win workers' loyalty in some work settings.

Braverman's notion of control has also been reconstructed by Edwards (1979), who identifies the emergence of three historically successive forms of control: *simple* control arising in small scale enterprises by direct supervision of the entrepreneur: *technical*, through the building of control into technological advances, such as the assembly line which regulated the speed of production; and *bureaucratic*, whereby rules define how the work is performed and govern the sanctions to be applied.

Following on from Braverman's omission of the arena of resistance and struggle, some writers have also criticised his conception of *management*. Salaman (1982), for instance, has argued that Braverman operates with a fairly simplistic functionalist notion of management as omniscient and conspiratorial, responding only to the imperatives of capitalism and capital accumulation with little room for the participants to actively intervene in the process through resistance.

Likewise, Braverman's conception of *Taylorism as the logic of capitalism* has been criticised. Littler and Salaman (1982), for instance, have argued that Braverman assumed the relatively unproblematic imposition of Taylorist management methods. One example they cite is one of the companies where Taylorist reorganisations of the labour process were introduced, which was not overly happy with the results achieved because the decrease in the size of the workforce brought about by the reorganisation, had considerable implications for the returns from company-owned housing for the workers. Taylorism was not the only such form of management practice; others, such as Fordism, the Bedaux system and the human relations approach, could achieve the same ends. Hence Littler and Salaman (1982: 257) argue, 'there can be no theory of *the* labour process...the linkage between the logic of capitalist accumulation and the transformations of the labour process is an indirect and varying one.' What Braverman did was to identify one type of control of labour and assume it expressed *the* logic of capitalism.

Related to this perceived problem with Braverman's analysis are further problems with the *geographical specificity* of his account. Braverman tends to assume, it is claimed, that his account of the degradation of work in the United States can be generalised to the broad concept of capitalism in general. As a number of writers (e.g. Littler, 1982; Burawoy, 1985) have claimed, there are specific variations between capitalist societies in the labour process which must be understood in terms of the historical constellation of struggles in different countries. Littler (1982: 50) refers to this as the 'Ambrit' fallacy: the assumption that what occurred in the United States also occurred in Britain and other countries. However, specific historical circumstances, such as greater tradition of resistance in Britain, meant that the struggle worked out differently there and in other countries such as Japan. This shows the importance of detailed historical studies in examining the variations between capitalist societies. Perhaps in the context of Australia and New Zealand, for instance, the consolidation of unionism prior to the emergence of large corporations and the transition to monopoly capitalism (unlike the United States) has had a significant effect on the development of the labour process, including the manner in which skill has been expropriated. In Australia and New Zealand, insulated behind tariff walls, perhaps concessions to workers could be more easily passed on to consumers, thus affecting the competitive pressure to transform the labour process in the direction Braverman outlines. Such questions as these are partly examined later in this book.

Another major line of criticism of Braverman's work is the perceived *inadequacy of his historical detail*. Braverman took, as his benchmark of the skilled and unalienated worker, the nineteenth century artisan in the engineering industries, against whom he traced the process of deskilling. Subsequent analyses have shown this to be based on an over-romanticised notion of the work of such artisans when, in fact, considerable subordination to capital already existed (see Thompson, 1983; Littler, 1982; Wood, 1982). Early capitalism has proved not to be the haven of the craft worker that Braverman implies. Littler (1982), for instance, has stressed the importance of subcontracting with individual artisans responsible for the control of other workers.

Likewise, Beechey (1982) has criticised Braverman's account of the relationship between gender and the labour process on the grounds of the historical inadequacy of his account of pre-capitalist families, arguing that he:

> has a glorified conception of precapitalist society and of pre-industrial capitalism. His model for the preindustrial family is based upon the farm family which produced sufficient to meet its needs and, in which (he assumes) work was a 'natural function'. This 'Golden age' conception of the pre-industrial family is both romanticised and ahistorical. (Beechey, 1982: 58)

Further analysis of Braverman's concept of *technology* finally has refined usage somewhat. Braverman tended to accord primacy to Taylorist attempts to deskill workers in transformations of the labour process, regarding technological innovation as being of secondary importance and not necessarily connected with managerial strategies of control. Although he acknowledges the incorporation of control into the machine through the Babbage principle, subsequent analyses have examined this issue in considerable detail. Thompson (1983), for instance, in an extensive review of the role of technology in restructuring the labour process, perceives a general tendency to the deskilling of the workforce, particularly with the advance of automation. However, he argues that there is no technological inevitability about deskilling; it may lead to the creation of new skills or of responsible autonomy for groups of workers. Empirical instances have been uncovered (e.g. Child, 1985: 130) where there has often been a conscious decision to utilise new technology especially for the purposes of deskilling and job degradation, even when alternative technology, or other forms of work organisation exist which could be utilised effectively with that technology, (see also Winner, 1985). One area of innovation in which this has been effectively demonstrated by Noble (1984) is the introduction of machine tool technologies such as numerical control.

These then constitute the major lines of criticism to which Braverman's analyses have been subjected. Together they suggest, to a greater or lesser extent, that, however successful Braverman may have been in posing crucial questions about technology, the labour process and advanced capitalist societies in general, he was less successful in the answers he provided to those questions.

Post-Braverman

The Braverman legacy, however, rejuvenated sociological studies of work and led off in a number of directions as attempts were made to satisfactorily answer the questions posed by Braverman in the sense of being, as Littler (1982: 34) puts it, 'less rhetorical and better grounded'. All sorts of issues have been examined, spurred by the labour process debates, some of which are represented in this book. Littler (1987: 57) has recently identified three core directions in which labour process theory has developed. The first and earliest of these is 'questions about deskilling and the attempt to construct a satisfactory model of skill changes'. This has been an attempt to better answer the question about skill changes that Braverman identified. An example of this direction is the volume by Wood (1982).

The second direction is 'questions about labour markets and the attempt to construct a satisfactory model of capitalist labour markets'. This has involved various attempts to develop accounts of labour markets in terms of dimensions such as internal and external, primary and secondary (see Friedman, 1977a). Attention has also been focused on the role of the State developing the earlier work by Burawoy (1985) on the transformation of factory regimes (see Thompson, 1986). In the Australasian context the centralised mediating role of the arbitration system has been analysed in the volume by Bray and Taylor (1986).

The third direction is 'questions about managerial strategy and control' relating more to sociology *in* work. An interesting example is a recent book of readings by Knights et al. (1985) which demonstrates the penetration of labour process concerns on to the terrain of *job redesign*, a traditional concern of the sociology in work approaches aimed at humanising or enriching the experience of work. Such concerns were dismissed by Braverman as attempts to habituate the worker to degraded work. However, as the editors point out, because the job redesign literature has until now been 'infused by a managerialist or technicist orientation to work organisation ... based on theories of motivation that abstract the individual from history and the structure of social relations' (Knights et al., 1985: 2–3) does not mean it will not necessarily always be so. Instead they aim, via a debate on aspects of labour process theory, to 'understand and explain job redesign as a medium and outcome of political, economic and social forces that interset within the organisation and control of the labour process in late capitalist society' (Knights et al., 1985: 8).

In addition to these three strands identified by Littler there is a fourth key direction in which labour process studies have developed. A rich line of investigation has been the feminist analysis of the relationship between gender and work. Examples of this strand of development are Cockburn, (1983); Game and Pringle, (1983); O'Donnell, (1984); and Knights and Wilmott, (1985).

Other lines of investigation into the contemporary organisation of work are continuing to develop as the agenda set by the questions Braverman posed are being worked out. One example of a promising line of investigation is in the analysis of the emergence of occupational health and safety issues as an important industrial relations issue in the social relations of work. One example of this research is Willis (1986).

Summary of this work

The aim in this book has been to produce a collection of lively and informative articles which engage the labour process debate in its Braverman and post-Braverman forms. All of the core directions in which labour process analysis is developing are represented in this volume. Some of the articles are less sympathetic to such labour process concerns than others, but all are to a greater or lesser extent within that approach to the study of work in late-capitalist Australian and New Zealand societies. All are concrete, historically grounded case studies of the development of the labour process in these historically specific late-capitalist societies. Furthermore, all deal, though to a varying extent, with the interaction between workers, managers and technology. Overall, the aim of the book as a whole is to argue for the utility of labour process theory in understanding the development of the labour process in a wide variety of work settings.

In Chapter 2, Dawn Butler takes up the theme of the skill

implications of technological change in white-collar office work. She argues that, with the exception of private secretaries, the major implication of changes in the labour process around new technology, has been the deskilling of the office workers involved. Rosslyn Reed, in Chapter 3, examines several themes associated with technological change in the printing industry, arguing for an interpretive account of such change through emphasis on subjectivity and human agency. In Chapter 4, Judy Wajcman and Belinda Probert look at the labour process implications of a deregulated labour market in the case of outworkers in the computing field. They focus particularly on gender issues in relation to the sexual division of labour. Kerr Inkson and Peter Cammock in Chapter 5 on the meat-freezing industry in New Zealand examine issues of resistance and managerial control in reorganisation of the slaughtering labour process. They argue for an account of subjectivity to be built into analyses of both worker and management behaviour. The managerial strategy of 'quality circles' is analysed in Chapter 6 by Richard Dunford and Peter McGraw. Taking the case study of a manufacturing plant, they argue the need for labour process analysis to develop a more subtle, less deterministic account of the operation of management control in the labour process. Raelene Frances, in Chapter 7, analyses historically the transformation of the labour process in the clothing and boot industries, especially as it has been affected by technological change. She argues for the complexity of the manner in which the twin processes of control and resistance operate.

In Chapter 8 Jeanne Daly and Evan Willis examine the professional labour process and labour markets in the health industry. Taking the case of diagnostic imaging technology, such as x-rays, they argue that accounts of ideology and the State must be incorporated into the analysis of professional work. The introduction of Japanese management practices into the motor vehicle building industry is the focus of attention of Chapter 9 by Roger Wilkinson. There he considers management practices and skill implications in the context of industry restructuring. In Chapter 10 Ern Reeders analyses the fast food industry, an important area of what has become known as the secondary labour market. He argues that skilled, satisfying work is designed out of the labour process as part of the managerial strategy of intensification of labour. Len Palmer, in Chapter 11 demonstrates the utility of labour process analysis to technological change in the telecommunications industry. He considers the effects of the interaction of such changes with job redesign on the skill levels of Telecom workers. Finally, John Mathews examines the issue of what a 'labour process policy' for trade unions might look like in Chapter 12. He argues that technological change is about *choices* and need not be approached fatalistically in a technologically determinist way by workers in Australia and New Zealand. Instead, such labour process analysis allows for the development of practical strategies by such workers and their collective organisations which constitutes a form of empowerment in influencing the sort of work in which most of us will be involved in the future.

One final point is a methodological one, an argument for the importance of the case study approach in understanding and analysing the labour process in particular and the subject matter of the social sciences in general. The advantages of grounded historical accuracy and richness far outweigh the disadvantages of generalisation. Such a methodology provides the best opportunity to counteract the ahistorical and/or empiricist direction in which much of sociology in particular has traditionally developed.

2 Secretarial skills and office technology

DAWN BUTLER

In the 1970s and 1980s, secretarial work has undergone great change. With the introduction of word processing technology, new jobs and new skills are being created, previous jobs are changing and there is some degree of displacement of secretarial workers in underemployment and even unemployment. In Braverman's words, office work is facing an 'incipient upheaval . . . We are at the beginning of a revolution' (1974: 341).

However, despite the fact that offices are changing significantly and that so many women (up to one third) are at some point in their lives involved in secretarial work, very little research has been done on secretarial work by sociologists of work or by feminists. The handful of studies which have been undertaken have been comparatively small-scale, and have tended to concentrate only on the social relations of the job (see McNally, 1979; and Benet, 1972). Other important work in the area, such as that done by Glenn and Feldberg (1979) and Barker and Downing (1980), while advancing the theoretical debate, is not based on detailed empirical work. The present case study conducted in Brisbane, Queensland, is put forward as a contribution towards understanding this important area of women's work, within a framework of feminism and the labour process debate.

In the context of the changes which are occurring in the modern secretarial labour process, two major questions need to be asked. Firstly, can it be said that the changing systems of work organisation enhance management *control* over secretarial workers? Secondly, does Braverman's *deskilling* (or work degradation) thesis provide an appropriate explanation of the transformations which are occurring in secretarial work? To answer these questions, this chapter will begin by briefly describing the modern office labour process and its development. It will then consider issues of management control and worker consent and contestation within the secretarial labour process. Finally, the chapter will consider the issues of skill and deskilling in relation to secretarial work.

The study

In this case study of the secretarial labour process in Brisbane in 1982, several research methods, both qualitative and quantitative, were used. These included the administration of a questionnaire with closed- and open-ended questions (900 questionnaires were distributed, with a response rate of 73 per cent), private in-depth interviews (29 interviews with secretarial workers and ten with managers and unionists) and two group workshops (each with 25 participants).

The questionnaire was distributed to secretarial workers in four types of organisations: insurance companies, legal companies, academic institutions and the public service. A fifth category, a general one, also emerged. In each category there were more than 100 responses. As well as covering workers in different types of organisation, the survey covered the full range of levels of secretarial workers: private secretaries, stenographers, clerk–typists and pool typists or word processing operators in centres.

The modern office labour process

The fact that almost all secretarial work is performed by women is not merely coincidental; secretarial positions are not simply 'empty places', 99.7 per cent of which (in the Brisbane study) happen to be filled by women. The way the job has developed, the nature of the tasks performed and the nature of the relationship between the secretary and her (usually male) boss all need to be understood in terms of gender as well as capitalist social relations.

Women entered office work at the time of industrial capitalist expansion. Previously it was men who had done the work of personal assistants to the owners, copying out documents, doing shorthand, bookkeeping and clerical work of all sorts. The expansion of capital brought with it the rapid growth of the office. As part of this growth, new technological developments included the telephone, the telegraph and the typewriter. Women were brought into offices to operate the new technology and to do other simple and routine tasks (which came to be seen as 'women's work'), while many male clerks moved up into lower and middle management positions.

There was some degree of resistance by the male clerks to women being employed in offices at all, especially as shorthand writers. However, once women combined the two skills of shorthand and typewriting, their positions in offices were assured. Thus two distinct occupational hierarchies developed in offices, divided along gender lines. In short, the office hierarchy was transformed as part of the growth of capitalism, and gender relations were integral to this transformation.

As well as gender divisions, there has also developed another significant division in offices, that between private secretaries and pool workers. On the one hand, in the traditional form of office organisation, often termed 'the social office', various managers each have a private secretary or access to a secretary, to tend to their business and personal needs in the office. These secretarial workers operate in a highly personalised gender-based relationship with their bosses, aptly summed up in the term 'office wife'. In this situation, Edwards' (1979) notion of 'simple control' is directly relevant: these workers are subject to the unstructured and often arbitrary power of their bosses, and refusal to perform any tasks is difficult.

On the other hand, there are other secretarial workers (typists and stenographers) who are organised into pools and have little direct personal contact with those they serve. The objective conditions of work for this 'pool proletariat' (Benet, 1972: 137) form the material basis for a very different set of social relations from those of private secretaries. The earliest pools were established at the turn of the century during the Taylorist drives for scientific management and efficiency. Conceived of in terms of direct control over the labour process, pools have always offered greater potential for work monitoring. While it is often assumed that pool work is a preparation for the position of private secretary, the findings of the Brisbane study suggest that the two groups of secretarial workers are increasingly separating and polarising, a process further advanced by modern office systems.

The most significant feature in the transformation of the office labour process in recent years has been the introduction of word processing systems. With the increasing use of word processing, the social office of previous years has become less widespread. Office organisation is moving towards a 'total system' designed to maximise the capacities of the new technology. This has brought a diminishing number of private secretaries on the one hand and, on the other hand, an increasing concentration of workers in 'word processing centres' or pools.

Looking at offices in general, it appears that office organisation around word processors has taken one of two forms: a centralised form whereby the machines and the operators are pooled; and a distributed form whereby 'the machine is brought to the operator rather than the other way around' (Reinecke, 1982: 155). The latter form, in most cases, refers to private secretaries, who are using word processors as a more advanced 'tool' than their traditional typewriters.

The centralized form of organisation has resulted in major changes in both the nature of secretarial work and in office social relations. In centralised offices, secretarial work has come to be divided into two parts: typing and administration. Typically the bulk of the typing work is done in the 'word processing centre' and most of the administrative tasks are performed in the 'administrative support centre', staffed by 'administrative assistants' (AAs), each of whom serves several 'principals'. The AAs handle the other secretarial functions such as filing, answering the telephone and making appointments. They do very little typing, except for some of the 'unique' correspondence typing. Thus, in centralised offices, the traditional secretarial function is fragmented and different secretarial workers specialise in different fractions of the job. Accordingly, the social relations of secretarial work have altered profoundly.

In several of the Brisbane organisations surveyed, work is forwarded to workers in the 'centre' or pool, either by written request or by recorded dictation (not by personal dictation, which is taken down in shorthand). Telephone or dictaphone dictation is encouraged by the equipment suppliers as being the most efficient way of operating. In an advanced centralised dictation system, recorded dictation can be automatically fed to an operator as soon as her 'work station' becomes free. There is an 'endless loop' system which can allocate work to operators. From the operator's point of view, she remains continuously plugged in to the system without any idea of how much work lies ahead (CSE Microelectronics Group, 1980: 48–9). This system of dictation control, together with the other monitoring facilities built into word processors, can provide management with almost total control of these workers, though it was found that few of the Brisbane organisations actually made use of such dictation systems.

The organisation of the secretarial labour process along centralised lines is not unlike the production process in a factory, as was evident in several Brisbane organisations. In insurance companies or legal offices, for example, letters may be 'originated' or conceived of by clerks, whose jobs have also been routinised to the degree where their task consists of simply selecting the appropriate clauses or paragraphs to make up a letter. Using word processors or 'mag cards', secretarial workers then type the appropriate clause numbers to produce the document. The process can take place at great speed and it is simplified to the degree that the likelihood of errors is minimised. Such a work design actually removes a large part of the production of documents from the typist to the machine, a shift which in Braverman's terms 'degrades' or deskills secretarial workers. This system has also significantly reduced the number of secretarial jobs. Insurance companies, in particular, have already dramatically cut back their numbers of secretarial workers. In the Brisbane survey, one manager commented: 'Five years ago, there were more than 100 secretaries. But now, in the whole of Queensland, there is only the secretary to the State Manager, and 10 girls [sic] in the Word Processing Centre. The rest have all been replaced by flashing lights.'

Not all modern offices use centralised systems. The alternative form of office organisation distributes word processors amongst secretaries throughout the organisation. It would appear that this 'distributed' approach is more popular with both executives and secretaries (Reinecke, 1982: 155), though some of the Brisbane managers maintain that productivity suffers when the machines are decentralised. Despite the questionable effects of centralised organisation, it is evident that larger firms in Australia are opting for centralisation (Economic Research Unit, 1980: 40).

As well as fragmenting secretarial work and reducing the numbers of secretarial jobs, the use of technology also offers management greater opportunities for control. For example, several computer companies advertise a type of 'management statistic program' to be used in conjunction with their word processors. This enables management to key into a control screen and get an exact picture of the office production process at its various points. The monitoring system gives precise information about which work stations and printers are being used at any particular time, which operators are using them, when they start and stop, which document they are working on, whether they are editing the document or keying it in, how many lines have been typed, how many errors have been made, how many revisions made, and how much central processing time they have used. Such a degree of control would indeed have been applauded by Taylor himself! And this possibility does not go unnoticed by the workers; many operators are aware of the degree of surveillance which is possible through the machines, and some have made reference to 'the spy in the typewriter' (Reinecke, 1982: 164).

From the Brisbane survey, however, it appeared that the technical possibilities for monitoring and controlling work are not generally used to the full. Some organisations do monitor output and pay wages accordingly. Others used to do so but have stopped. As one manager stated: 'Originally typists were rated on the number of words they could type, and they had to get through so many a day. But we've scrubbed all that because there was no need for such sophisticated productivity checks.' Whether these checks are used or not, it is clear that management control is essentially *built into* the modern office labour process and management has some choice about the degree of control exercised, influenced also by the extent of worker contestation and consent.

In the office labour process, different forms of control can be seen: the simple control of secretaries by bosses, the direct control of pool workers by supervisors and the actual and potential controls of workers through machines. However, controls are not simply imposed unproblematically by management and accepted passively by workers. In the case of secretarial workers as well as others, it is necessary to consider issues of worker consent and contestation in the shaping of the labour process.

Following Burawoy (1979), the theory of hegemony can be applied to the secretarial labour process to understand the ways in which workers consent to and contest the conditions of their work. Like other workers, secretarial workers play an active part in creating and maintaining office social relations. In the gender-based ideology prevalent in offices, there are specific norms of appearance, dress and behaviour, all of which are aspects of personal control over secretarial workers. The workers themselves often uphold and even generate these norms. In the Brisbane study it was found that most secretaries collude with their 'office wife' roles, sometimes with approval, sometimes with apathy and sometimes with resentment. However, even where there is resistance, this is rarely contestational and, when faced with ultimate dissatisfaction, private secretaries are more likely to leave the organisation than contest within it. This is not to say that private secretaries are passive and simply reactive. In many cases they are able to manipulate the personal relationships of the job to their own advantage. Situated within the corridors of power, with access to high level information, many secretaries enjoy a good deal of informal power. Seated at the door to his office and keeping his diary, the secretary is able to block or filter access to the boss and to organise and prioritise both information and people.

In contrast to private secretaries, for the 'pool proletariat', the social relations of office work are quite different and provide the basis for different forms of consent and contestation. Pool workers work in close physical proximity to each other, in physical isolation from the rest of the organisation and under the control of a supervisor. In these conditions, a type of camaraderie develops which is quite different from the isolation of private secretaries. Partly as a response to work pressure and partly as an expression of group identity, a feature in many pools is peer control of the pace of work. Workers may set themselves and each other a mutually understood and comfortable pace of work and they are able to communicate their disapproval to any who may break solidarity. In contrast to private secretarial work, collaborative practices such as work pacing, work sharing and a more positive attitude to unions were significant features of many Brisbane pools. However, even in those cases where worker response to pool conditions was negative, this was seldom developed into overt contestation. As with private secretaries, the main alternative for pool workers in times of conflict is to leave.

In terms of their influence on the developing labour process, Brisbane secretarial workers were seldom consulted on the introduction of new technology and most were overwhelmingly approving anyway. The majority of the Brisbane respondents (65 per cent) said that they had been consulted 'very little' or 'not at all' when word processors were introduced. This low level of consultation has clearly given rise to the clerical and insurance unions' regular demand that they and the workers affected be consulted. However, in the remaining 35 per cent of cases where 'consultation' had occurred, it tended to be simply a matter of informing the secretarial staff of decisions already made by management. In organisations which had already introduced word processors, 84 per cent of the secretarial workers approved of them. And in those organisations where word processors had not yet been introduced, the approval was even greater (86 per cent). However, a few respondents did express some reservations in interviews: they feared that they would be displaced by the machines, or they feared the effects of using the machines ('make them dull and conversationally almost stupid') and the extra workload that would be caused by the machines ('old duties as well as new ones').

Despite the overwhelmingly positive attitudes of the Brisbane respondents to word processors in general, many workers were prepared to express a specific concern about health problems arising from their use. Equally strongly, several of the managers interviewed dismissed these concerns as 'a figment of the imagination', 'a load of rubbish', or 'bees in bonnets'. Nevertheless, of those respondents working with word processors, half considered that word processing does cause health problems, while half denied the link or said they did not know. The problems most often referred to were eye problems and stress. In interpreting this, however, it should be borne in mind that the survey was conducted in 1982, and it could be speculated that with subsequent wider and more intense use of the new technology, complaints of repetitive strain injury might feature more prominently.

Thus, while there is a general attitude of the inexorable 'march of technology', many workers have identified problems through their and their fellow workers' experiences relating to the machines. However, at this stage the workers have largely consented to the changes; it remains to be seen whether their reservations will in time take shape in articulated opposition.

This section has concentrated on changing forms of office organisation, especially in relation to the use of modern office technology. As Braverman argues, the introduction of office technology is only partly related to issues of efficiency and productivity; it is also related to issues of control over workers. Having considered issues of management control and worker consent and contestation, this chapter will now look at the issues of skill and deskilling in relation to secretarial work.

Changes in secretarial skills

Labour process literature contains an extensive debate around the notion of 'skill'. In the limited space available here, suffice to say that the notion of 'skill' has at least three components, not all of which need apply before the term is used. (For a fuller discussion on the notion of skill and its application in the secretarial context, see Butler, 1986.) Firstly, the *worker* may be skilled; s/he may be able to exercise knowledge which incorporates both conception and execution of a particular set of tasks as well as having some degree of control over the labour process. Secondly, skill may be a feature of the *job*. This component is to be distinguished from the concept of the skilled worker because the skills which a worker may have may not match the skills required by the job. For example, a skilled stenographer may be asked to operate a word processor, a situation where the worker may have skills which are not required, and not have skills which are required. This notion of skill is broader and is more useful when considering whether the skill requirements of jobs are changing especially with regard to the issue of deskilling. The third component of skill is its *social or political definition*, where a certain skilled or unskilled status may be attributed to a job without this necessarily reflecting the actual skilled nature of the task, or where workers have been able to negotiate one or the other of the labels together with corresponding wages.

Combining these components of skill, I would argue that secretarial workers are skilled workers, in skilled jobs, but that they enjoy only limited negotiated status as such. Definitions of skill are not static categories. Just as the labour process is itself continually changing, so there are also ongoing negotiations around the organisation and definition of work. In these processes of change and negotiation, ideology may be a powerful determinant of the status of the skill.

Keyboarding skills provide an example of gender influences on skill definition. The keyboarding skill performed by a word processor operator or a typist (usually a woman) is undoubtedly comparable with the keyboarding skill exercised by a compositor in the printing industry (usually a man), in terms of the combination of conception and execution, and in control over the labour process. However, in terms of social definition and negotiation of job status, the two jobs are not accorded the same level of skill. Cockburn (1983) discusses the fact that compositors have thus far managed to maintain their status as skilled workers even though she quotes one compositor describing his work as 'an idiot system . . . The brainpower aspect [has gone] clean out the window' (1983: 115–16). Conversely, typists are not conventionally regarded as highly skilled workers, either in ordinary discourse or in terms of union negotiation. From the Brisbane survey it is clear that many secretarial workers, especially those with word processing skills, sorely feel their lack of skill recognition.

Looking at secretarial work in terms of skills, it is not difficult to argue that secretarial work involves the conception and execution of a set of tasks as well as some control over the product. Typing, for example, requires keyboarding dexterity, speed and accuracy. In addition it also requires knowledge of display conventions, including how and when these are used. Shorthand writing, another secretarial skill, involves the ability to use symbolic outlines to record the spoken word. To write high-speed shorthand and then transcribe it accurately requires manual dexterity and considerable competence in comprehension and language skills, as well as a good memory. Both typing and shorthand skills take months or years to acquire and require ongoing practice to maintain. In the Brisbane survey, secretarial workers reported high typing and shorthand speeds: 75 per cent of the respondents could type at speeds of 60 words per minute or faster; 25 per cent of them had shorthand speeds of 120 words per minute or more (i.e., verbatim speed). Such speeds reflect high levels of skill competency. And, of course, in addition to the skills of typing and shorthand, most secretarial workers have a range of other office skills, such as filing, duplicating , switchboard operation and so on.

In the context of the changing office organisation around the new technology, the demand for secretarial skills is clearly changing. As typewriters are being replaced by word processors, there is a demand for an entirely new skill. Consequently, certain aspects of the typing skill have become obsolete since word processors have some of the previous typing skills built into them. For example, being able to type tabulations would traditionally have been regarded as an important part of a typist's skill, but when using a word processor the machine can produce tabulations at the touch of a button. Cooley refers to such a transference of skill:

> the knowledge [and skill] which previously existed in the consciousness of the [designer], which was his or her knowledge . . . has now been extracted from them. It has been absorbed and objectivised into the machine through the intervention of the computer. (1980: 8)

Inevitably the worker's skill becomes lost through lack of practice and at the same time becomes technically obsolete.

Without doubt, the ability to operate a word processor is a significant *new* secretarial skill. Firstly, in terms of the criteria for skill referred to earlier, being able to operate a word processor represents a change of skill in the worker herself. Secondly, there is also a change in the job itself as office systems change. Thirdly, however, in terms of the political and social definition of word processing as a skill, there is a lag on the part of employers who in many cases do not recognise word processing as an extra skill which deserves extra remuneration. In the Brisbane survey there was general agreement that complete competence in word processing may take a year or more to acquire. On the other hand, equipment suppliers usually claim that the skill is quick and easy to attain, and accordingly offer short three-day courses, which many operators regard as insufficient. The inadequate training, together with employers' unrealistic expectations of doubled and trebled productivity, also devalues word processing as a skill. Inadequate training often causes intense pressure and anxiety for operators in the first few months of operating. Despite considerable effort, some respondents feel that they are blamed for inefficiencies which could have been avoided by more adequate training and a more careful process of implementation and transition.

While it is true that new office systems have provided opportunities to gain the new skill of word processing, it is also true that opportunities to use other skills such as shorthand have disappeared in the pool situation, which is a development much regretted by those secretarial workers who have the skill. However, it is only [illegible]orkers in word processing centres with recorded dictation [illegible] who are denied all opportunity to use of their shorthand skill; for private secretaries, shorthand continues to be a valued skill alongside the newer skill of word processing.

A further microprocessing facility in the office which bypasses the need for both shorthand and typing is the Optical Character Recognition (OCR) machine, which can 'read' a typed document and transmit it directly to a central processor which 'types' a new document. And, in the longer term, the final displacement of shorthand could come from 'voice input'. At present the technology for voice or speech input is relatively underdeveloped, but the potential already exists.

Conclusion: Deskilling and/or enskilling?

In the context of the modern office labour process, the question needs to be considered whether the restructuring of the secretarial labour process deskills or enskills secretarial workers or, possibly, whether *both* processes are occurring in the same worker, or in the sense that some secretarial workers are being deskilled while others are being enskilled.

Clearly, regarding word processing equipment itself, it needs to be said that the equipment *per se* does not necessarily deskill or enskill jobs or workers. Rather, the impact of the new equipment needs to be assessed more broadly in terms of management strategy, worker resistance and office restructuring around its use.

Proponents of the upgrading (or enskilling) thesis argue that the new technology has mechanised many of the routine and alienating tasks previously performed by secretarial workers and has thus left them with more time and opportunity to concentrate on the more interesting and skilled tasks. Further, it is argued that those secretaries who have achieved proficiency in the new skills now have greater rewards and security in their jobs, have better prospects for promotion and enjoy a more 'humanised' environment.

Prominent support for the upgrading thesis was given by the Committee of Inquiry into Technological Change in Australia (CITCA) (1980), which quotes a statement from the Public Service Arbitrator claiming that there is 'evidence of heightened work interest' when using the new technology. Further, CITCA states that, in the private sector, word processor operators' salaries are generally higher than typists' salaries and that word processor operators 'are considered to hold elite positions in relation to typists and typist–clerks' (1980: 56). Less glowingly, the report acknowledges that word processor operators' career paths are short, perhaps even shorter than those of typists 'who can aspire to secretarial positions or occupations requiring skills in human relations' (1980: 56). However, CITCA concludes its section on word processors with the statement that: 'On balance, skills and responsibilities of typing staff are likely to be upgraded as a result of introducing word processors'. (Vol. 1, 1980: 57)

In contrast to the upgrading thesis, I wish to conclude with the claim that the majority of secretarial workers (in word processing centres and administrative support centres) are being deskilled by changes in the labour process around the new technology, even those workers who have acquired the new skill of operating a word processor. A clear exception is the private secretaries who may operate a word processor within the context of their previous functions and social relations.

Braverman's criteria for assessing the 'degradation' of work are useful here. The process of degradation begins with the division of labour, primarily by separating conception from execution. This division then opens the way for the progressive fragmentation, rationalisation and mechanisation of jobs. Through this process, complex labour becomes collective labour with the inevitable effect that individual workers have very little, if any, control over the total labour process, and become increasingly alienated.

One of the earliest attempts to substantiate this deskilling thesis in the clerical context was made by Glenn and Feldberg (1979). In their examination of the 'objective conditions' of clerical workers, they argue that these workers have been 'proletarianised' by a range of factors: by the new office organisation, by the introduction of office technology and by bringing scientific management into the office. They conclude that managers' increased control of the work process affects workers 'psychologically, economically and politically—as individuals and as members of a class' (1979: 71). Insofar as word processor operators in a centre or pool are seated for the large part of their working day in front of a machine, physically separated from other workers, being actually (or potentially) monitored in precise detail, and being fed dictation continuously through a centralised dictation system, these workers could perhaps be described (as they are by Glenn and Feldberg) as a 'collection of bodies mechanically related to a set of materials and sustained in motion by external force' (1979: 71). However while this description is to some extent substantiated by the Brisbane case study, it is clearly an overstatement. To call these workers 'inert' would be to ignore them as 'living agents' who continually contest and consent to the capitalist labour process and its social nature. Nevertheless, it remains true that under these conditions operators have comparatively little control over their labour process. Nor do they have any opportunity to gain knowledge of the wider production process. It would seem justified, then, to describe these workers as having been deskilled.

Private secretaries, on the other hand, have to date escaped such a degree of deskilling. These workers are being increasingly polarised from other secretarial workers; indeed, it has even been claimed that they are being 'professionalised' by word processing (SPRU Report, 1982: 74). The term 'professionalisation' is clearly being used loosely in this context since in no sense would these workers meet the usual set of criteria for professionalism. However, the term is obviously intended to indicate an enskilling or upgrading effect on private secretarial workers. While continuing to work within the more traditional personal relations of the boss–private secretary relation, these workers are not physically separated and situated in depersonalised pools. They are not ordinarily subjected to the same degree of monitoring. They have continued to utilize their previous skills (including shorthand, for many) and have, in addition, gained the extra skill of operating a word processor as an extra tool in the performance of their global secretarial function. These workers are also more likely to perform a wider range of tasks which may encourage them to utilise more of the interesting and challenging functions of the word processor. They are also likely to have a broader understanding of the work process in the organisation. Presumably these are the workers who are referred by CITCA as experiencing their work as 'creative' and challenging. To the extent that these secretarial workers are given more opportunities to develop their skills, they could be said to have been enskilled and upgraded.

Overall, it is clear that the issue of deskilling and/or enskilling of secretarial workers is not only a question of changes in the workers' skills. It must also be seen as a question of the way in which the job is differently organised as changes occur in the labour process. And here, as I have shown, issues of management control need to be considered, together with worker consent and contestation in the shaping of the labour process.

3 From hot metal to cold type printing technology

ROSSLYN REED

The earliest post-Braverman studies of technical change conducted within a labour process framework emphasised aspects of the process of technical change which validated the Braverman thesis (e.g. Zimbalist, 1979). Later attempts stressed market factors (both product and labour markets) and the effects of worker resistance in structuring outcomes of employer initiatives in technical change (e.g. Wood, 1982; Wilkinson, 1983; Knights et al., 1985). The emphasis on technology as a means of deskilling and degrading work which characterised the post-Braverman analyses of the labour process was a significant response to assertions that industrialisation and post-industrialism promoted increasingly skilled and intrinsically rewarding job design, which has contributed significantly to the emergence of a paradigm within which researchers are increasingly committed to concrete, historically informed analyses of changes in the organisation and content of work.

Printing and newspaper publishing in particular is an industry with craft traditions of labour process control. Recent technological developments, particularly in electronic typesetting technologies which have the potential to undermine those traditions, have been resisted by print workers. Strikes and lockouts throughout the 1970s and early 1980s have characterised attempts to introduce the new technology in the United States and the United Kingdom as well as in Australia.

Zimbalist (1979) emphasises the strategies of employers surrounding technological change which led to the loss of craft control over the labour process through confrontation and strike action. In a detailed analysis of the failure to reach agreement on technological change in the United Kingdom in the 1970s, Martin (1981) notes the divisions between employers and workers and the failure of State intervention to produce agreement. Both capitalist strategies and divisions within the working class, particularly based on the relationship between gender and the labour process, are emphasised by Cockburn (1983) in her dialectical approach to the contradictions posed for men and women workers by printing technology. She argues for a critique of technology, an examination of the political uses of skill, a redefinition of work and a single union for all workers in print (Cockburn, 1983: 223–33). Comparing United States, United Kingdom and Australian experience, Griffin (1984) concludes that the British print union movement has succeeded in buying time in which to achieve industrial unionism and retain control of print technology (1984: 59–60)

Subsequent confrontations between print workers, police and security guards at 'fortress Wapping', the new technology plant of Rupert Murdoch's News International organisation in London, became 'news items' in their own right. These developments tend to support the arguments of Zimbalist (1979), rather than the predictions of Griffin (1984). There have been suggestions that this is the 'Australian way' of dealing with industrial relations conflict. Yet new print technologies have been introduced in the Melbourne newspaper industry with a minimal degree of disruption to production and that on the part of journalists who, it could be argued, were in a position to gain from the change. To date, print workers and their organisations have been able to ensure that retrenchments comparable with those in other countries were avoided. In doing so, they engaged in strategies of resistance through bargaining and negotiation which have a long history in the Melbourne newspaper industry.

The structure of this industry has, however, changed in recent years. Ownership of David Syme & Co. Limited at the time of writing was in a state of flux. The New South Wales-based Fairfax organisation currently wholly owns David Syme and Co. Limited, the organisation on which this study is centred. The takeover of the Herald and Weekly Times (HWT) group by Murdoch's News Limited was completed early in 1987. Questions concerning the structure of ownership and the financing of these takeovers are outside the scope of this paper, but they nevertheless raise questions about the future of industrial relations practices in the industry, in particular the practices of negotiation and conciliation. Further questions arise surrounding the extent to which existing and historically developed divisions within the industry are likely to weaken printers' resistance to future technological innovations. This chapter attempts to analyse the development of those practices and examines the way in which they have structured the processes of introducing electronic typesetting at the Syme organisation. In doing so, some account is taken of constraints on both management and workers in adopting conciliatory rather than confrontationist strategies.

Study of technological change in newspaper production provides the opportunity to explore a number of strands in labour process analysis. By adopting the case method, and with a shift in emphasis from a Marxist structural focus to a Weberian concern for subjectivity and the reintroduction of human agency through concepts such as social closure and rationality and meaning, an interpretive account of technological change as a social process is possible.

An interpretive account does not, however, ignore the struggle for control between capital and labour as documented by Zimbalist in the United States (1979). Neither does it ignore the part played by the State and industrial relations institutions in channelling that struggle into specific forms as shown by Martin (1981) for the United Kingdom. Relations between labour and capital and the form that struggles over the labour process take are, however, also shaped by the existence and nature of divisions within the labour force. These divisions result not only from the actions of capital in structuring the labour market to ensure control over the labour process (Edwards, 1979), but reflect historically developed strategies of social closure, including the use of exclusionary strategies against contending social and occupational groups within the labour market (Parkin, 1979). One example of such a strategy concerns women workers when occupational groups (such as male-dominated trades and crafts) attempt through a number of tactics, including collective bargaining and apprenticeship requirements, to maintain control over areas of production (Hartmann, 1976; Rubery, 1980).

Where occupations become strongly identified as 'men's work' there is the potential for technological change to form the basis for a redefinition of the nature of the labour process. As Cockburn's (1983) Marxist–feminist account of the introduction of computerised photocomposition in the United Kingdom has shown, analysis of the gendered nature of work extends explanations of technological change as a process of 'deskilling'. An interpretivist account further focuses on the meaning of work and the development of traditions, cultures and sub-cultures within organisations and occupational groups which shape and legitimate both managerial strategies to control the labour process and the form of resistance undertaken by workers. Consent to the relations of capitalist production may, then, be the outcome of a minimal meshing of interests rather than the more contrived outcome outlined by Burawoy (1979). Actions can be both substantively (instrumentally) rational, appearing even pragmatic to both actors and social analysts. At the same time, they embody a value rationality which gives meaning to the action and, in some instances, channels action in directions which are not altogether chosen by the actors. Such traditional ways of acting can, on the other hand, be more or less consciously upheld (Weber, 1947: 116; Brubaker, 1984). As printers attempt to retain control over an area of production claimed by right of tradition as well as more instrumental subsistence considerations, they legitimate their actions through a belief that 'letter assembly is our work'. In retaining control over 'letter assembly' they are confronted with redesigned jobs which bear little resemblance to the occupations with which they identify.

The remainder of this chapter will outline the context in which a Melbourne newspaper publishing organisation, David Syme and Co. Limited (publishers of *The Age*), between 1980 and 1983, introduced the new typesetting technology of 'cold type', based on a 1979 technology agreement with the Printing and Kindred Industries Union. The technical focus for the study is limited to the introduction of electronic keyboards and visual display terminals (VDTs) into the labour process of printers, journalists and telephonist–clerks. The size and nature of the Saturday *Age* and its relationship to the profitability of the organisation, the competitive market for classified advertising which existed alongside the oligopolistic market for newspapers and the divisions of skill and gender in the labour force and the external labour market as well as the influence of the founder, David Syme, on both the organisational culture and occupational sub-cultures within the organisation, all contributed to the structure of collective bargaining between print unions and newspaper publishers in Melbourne. An evaluation of the extent of the 'deskilling' of printers both before and after the change and the effect of the redesigned jobs on job satisfaction will follow. The chapter concludes with some comparisons with other societies to indicate the contribution of local historical structuring factors.

The study

The data for this analysis were gathered in 1984–85 from observation and informal discussion with the occupational groups on the job, analysis of documentary evidence in an informally administered company archive and Printing and Kindred Industries Union (PKIU) chapel meeting minutes. Structured job satisfaction interviews were conducted with a stratified random sample drawn from each of the occupational groups and informal and focused interviews with some members of management and trade union officials. To analyse job satisfaction, measures drawn from the job satisfaction literature were used, selected on the basis of detailed observation of the labour process and taking account of the sub-cultures which mediate the experience of work for different occupational groups.

The technology

Typesetting, initially the assembly of individual letters and subsequently the production of strips of text on the Linotype machine, is central to 'hot metal' methods of newspaper production. Type is assembled into a page and locked into a 'forme' or 'chase'. When proofreaders are satisfied with the accuracy of the page, a papier mache 'mat' mould is produced by obtaining an impression of the information contained in the page, from which is mechanically cast the curved, lead, stereotype plate to be mounted on the cylinders of rotary printing presses, from which newspapers are produced. (See Zimbalist, 1979: 104–17 for a more detailed description.)

From the fifteenth century, when Gutenberg invented movable type, to the late eighteenth century, developments in printing technology remained static (Marshall, 1983: 13). Nineteenth century developments centred on press technology. During this period a degree of specialisation entered the craft. From the late nineteenth century when Mergenthaler developed the Linotype machine, technical developments have centred on typesetting (Smith, 1980: 80; Marshall, 1983: 22). Despite innovation in teletypesetting (TTS) and automatic linecasting machines in the 1950s and optical character recognition (OCR) in the 1970s, no significant developments in composition occurred prior to the development of photocomposition and electronic typesetting and their integration in computerised photocomposition (Marshall, 1983: 30). Developments in TTS were, however, significant in signalling to print workers the direction of future developments.

Developments in electronic photocomposition are of greater significance for printing and newspaper publication than any development since the invention of movable type, because they alter processes of production and therefore potentially the division of labour at the point of origination of copy (Marshall, 1983: 65). They represent both a continuation of twentieth century technical developments in printing technology and a decisive discontinuity in the social relations of print. 'Cold type' is the name generally given to the process of computerised photocomposition. Copy is keyboarded directly to a computer from which it is produced as 'bromides'. These columns of text on photographic paper are assembled into a copy of the finished page by 'cut and paste'. This is used to produce a negative for electronic photo-polymer platemaking. These flexible plates are then attached to the rotary printing presses. This process has been developed with the potential to bypass the processes of 'hot metal' (typesetting, page composition, stereotyping) altogether.

Theoretically newspapers *can* now be produced with a few skilled computer programmers, journalists and typists at one end, and despatch workers at the other. In practice this is not generally the case for Australian metropolitan newspapers, partly because of the proprietors' high investment in letterpress machinery and partly because of the specific resistance on the part of print workers. At the Syme organisation, the size of the Saturday edition of *The Age* also prevented early adoption of the new technology because a system which could handle the volume of classified advertising carried had not been developed.

The labour market

Computerised photocomposition and the associated equipment were introduced into an already historically segmented labour force. Distinct segments of the labour force and the labour market are represented by different trade unions: the Printing and Kindred Industries Union (PKIU) (printers), the Australian Journalists' Association (AJA) (journalists) and the Victorian Printers' Operatives Union (VPOU) (telephonist–clerks), which are the outcome of attempts on the part of occupational groups (most notably the craftsmen printers) within the division of labour in newspaper publishing to secure a degree of control over their inter- and intra-class relationships with both employers and competing social groups within the industry. Prior to the development of electronic typesetting, Melbourne newspaper printers, like their counterparts in the United Kingdom and the United States (Cockburn, 1983; Hartmann, 1976), had succeeded in effecting social closure (Parkin, 1979; Murphy, 1986) through historically developed trade union strategies to exclude women and less skilled workers from their craft in order to maintain and extend the rewards, privileges and controls over work claimed as the domain of male trade/craft workers.

Trade union histories (Hagan, 1966; Fitzgerald, 1967) reveal the primacy accorded to the goal of regulation of the labour market through apprenticeship from the inception of the Melbourne Typographical Society (MTS) in 1867 and continuing throughout its subsequent history within wider industry trade union structures in the Printing Industries Employees Union of Australia (PIEUA) from

1921 and the PKIU from 1966. Although the MTS contested and negotiated with proprietors over wage (piece) rates and working conditions, disputes were not pursued during periods of economic downturn, particularly in the nineteenth century, if strike action would destroy the union. The long-held goal of regulation of entry to the trade and the exclusion from union ranks of unskilled workers was achieved in 1928 with printing being one of the first industries to come under the Apprenticeship Commission (Fitzgerald, 1967: 146).

Apprenticeship is crucial to the regulation of the industry in the interests of skilled workers against unskilled labour (including women). Not only does it mobilise State support and provide legal sanctions, but the structures of apprenticeship also operate to legitimate the higher status, wages and labour process control exercised by trades/craftsmen. Limited access through a period of 'time' operation on Linotype machines prior to entry to the 'piece battery' reinforced commitment to the occupational structure and culture. Through apprenticeship, young workers, who may be the sons, nephews or grandsons of printers or other tradesmen, learn the appropriate behaviour of their occupational group and the maintenance of social distance from other trades and less skilled workers. They also learn the obverse: the manner in which independence is exercised from management to maintain control over work (Penn, 1985: 4–13). They learn the vocabulary, myths, traditions and the internal discipline (i.e. the sub-culture) of their occupational community. They learn, at least within the Melbourne printing industry, that their ability to shut down the production of the newspaper is not to be exercised lightly. Control is equally exercised through the successful achievement, reproduced daily against all obstacles, of a high quality product. This control, since the turn of the present century, has centred on the piece rate system of Linotype machine operation which has operated simultaneously to the advantage of an elite within the craft in providing high wages and controls over work and to management in holding back production deadlines as late as possible (Hagan, 1966: 335). Craft control is therefore characterised by a high degree of consent by these workers to the exploitation of their labour in the interests of management.

For rank and file printing workers, 'the chapel is the union'. It is most significantly at the level of house organisation or 'chapel' that contestation by printers and other newspaper workers takes place. Direct bargaining over wages and working conditions under the leadership of the 'Father of the Chapel' (FOC) has continued alongside developments in branch and centralised unionism and the institutionalised procedures of conciliation and arbitration. The FOC combines in his duties the internal disciplinary functions of guild 'father' and the representation of group interests to, and on behalf of, workers characteristic of a shop steward (e.g. Fitzgerald, 1967: 125) with emphasis shifting from the former to the latter over time. Prior to recent technological changes, these channels of negotiation and contestation maintained craft traditions of control over the labour process and higher wage rates than in commercial printing.

Consistent with their strategies to exclude the less skilled workers of the print industry from the rights and privileges they claimed by craft tradition, printers gave little support to the organisation of the less skilled workers who posed a potential threat to their status, control and rewards. Reluctantly, the MTS assisted in the formation of the Printing Trade General Workers' Union in 1910 (Hagan, 1966: 153), subsequently the VPOU (Fitzgerald, 1967: 120). Since 1919, clerks and, since the late 1940s, telephonist–clerks employed in the Melbourne newspaper industry, have been members of the VPOU. Despite amalgamations in the direction of an industrial union, the VPOU remains an independent Victorian union separate from the PKIU, engaging in direct negotiation of private agreements with the Metropolitan Daily Newspapers Association of Melbourne, attempting to win for its heterogeneous membership a share of the rewards and privileges which accrue to members of the PKIU on the basis of craft and skill.

As an intermediate occupational group, journalists have been concerned to limit the power of proprietors to exploit their labour and to improve the status of their occupation. The actions of printers had little influence on the development of trade unionism among journalists, except as an example of the achievements of collective action (Lloyd, 1985: 22). Following failure to organise along more traditional 'professional' lines, the AJA was formed in 1910 as an industrial union registered under the Commonwealth *Conciliation and Arbitration Act* to obtain better wages and working conditions (Lloyd, 1985: 64). The Victorian branch of the AJA engages in collective bargaining with the Melbourne newspaper publishers' association. In addition, the AJA has also been concerned with attempts to achieve 'professional' status through the development of a system of cadetship, a code of ethics and structures of internal discipline.

Over time, then, these three occupational groups have, in varying degrees, engaged in and contributed to the development of historically structured procedures of collective bargaining through negotiated industrial agreements with the proprietors' association, with recourse to arbitrated settlements in the event of deadlock. At the same time they have developed distinct sub-cultures of work which, particularly in the case of printers and journalists, legitimate their position in the division of labour in newspaper publishing. The preference for collective bargaining is not, however, the result of action on the part of the unions alone. It is favoured also by the Melbourne newspaper proprietors in general and David Syme and Co. Limited in particular.

Organisational culture

The traditions and organisational culture of David Syme and Co. Limited derive from the origins of *The Age*. Their contemporary salience for a study of technological change results from a revival and redefinition of the traditions of David Syme by his great-grandson, C. R. (Ranald) Macdonald and his high-profile editor, (the late) Graham Perkin, in the late 1960s and 1970s, which both created pressures for the facilitated management attempts to introduce new technology. Investigative reporting was a key feature in changing the editorial stance of the paper from its then conservative attitude to a liberal approach in order to attract a new readership and hence a greater share of advertising. By linking this editorial strategy to the history of the newspaper (see Macdonald, 1982: 27), Macdonald emphasised the shared 'heritage' of the printers and the organisation.

The period between December 1854 and May 1856, during which *The Age* was published by a co-operative or 'co-partnery', has promoted a continuing identification with the organisation on the part of printing craftsmen. *The Age* co-operative's support for the Eureka rebellion, Ebenezer Syme's participation in the co-operative and David Syme's identification with the printing craft (Sayers, 1965: 10, 30–1, 266), combined with his liberal–humanist newspaper policies such as land reform, protection for local manufacturing and opposition to 'sweating' labour which coincided with the interests of local capital and skilled workers in Victoria in the nineteenth century (Hagan, 1983: 31–3), reinforced printer identification with *The Age* and the organisation. In his Daniel Mannix Memorial Lecture in 1981, Macdonald notes the historical identification of the printers with the organisation through their financial contribution to the expenses of an 1893 libel action: 'Unhappily, for modern newspaper publishers the days of such enthusiastic commitment to any single newspaper have probably long gone.' (Macdonald, 1982: 26).

From the death of Ebenezer in 1860 until his own death in 1908, David Syme was, in effect, in sole control of *The Age*. Throughout the nineteenth century *The Age* and the organisation were dominated by his liberal–humanist beliefs and values and the combined power of his eloquent pen and enigmatic presence. He appears to have conducted his business with a combination of entrepreneurial shrewdness and unswerving moral commitment (see Sayers, 1965). He was, in short, the material from which organisational myths and legends are made. The continuing Syme family association with the newspaper for more than a century reinforces the paternalism of print with notions of 'family' and liberal–humanist tradition surrounding organisational practices.

A significant factor in Macdonald's opposition to the first attempt by international media entrepreneur Rupert Murdoch to gain control over the HWT organisation in 1979 stemmed from his concern that this would alter the consensual approach to industrial relations by Melbourne newspaper proprietors (Souter, 1981: 573). Macdonald's dismay at the Fairfax purchase of HWT shares to counter the control bid by Murdoch reflected the significance attached to 'independence' in editorial comment (Souter, 1981: 578), a strategy related to advertising policy dating from David Syme's confrontation with merchant (British) capital in the nineteenth century.

Despite changes in company structure from 1943 and a partnership with John Fairfax and Sons from 1966, *The Age* remained under the control of David Syme's descendants until 1983. Their goal continued to be the publication of *The Age* as a 'quality' *independent* newspaper (Souter, 1981: 417–18). To this end, the company has sought to establish and maintain its dominance in the classified advertising market and expand its share of the display advertising market.

Collective bargaining and consensual industrial relations, then, can be seen to be the outcome of developments in trade unionism and organisational practice and culture. The organisational culture and the sub-cultures of the occupational groups mesh in a manner which produces ideological control over the labour process by the proprietors, while at the same time allowing a high degree of worker control over what is essentially a capitalist labour process. Given the primacy attached to classified advertising and the perishable nature of the product, this situation is particularly advantageous to management.

The product market

To date the Melbourne newspaper market has operated as an oligopoly. As the Norris Report (1981) demonstrated, *The Age* occupies a segment of the market, directing its editorial and news policy to a better educated, more affluent readership. Quality production and presentation, as well as independence in editorial comment, are all mobilised to this end. This strategy is directed towards the selling of both classified and display advertising space. Here the Melbourne newspaper publishers engage in open and determined competition.

Throughout the 1970s and the early 1980s, the revenue from advertising, and hence profitability for *The Age*, fluctuated according to the economic cycle. In 1983 the HWT attempted unsuccessfully through what has become known as 'The Herald War' to achieve dominance in real estate advertising (see Reed, 1987). Following the introduction of computerised photocomposition, profit in 1983/84 increased by 100.5 per cent and in 1984/85 by 259.8 per cent (David Syme and Co. Limited, *Annual Reports*, 1984: 5; 1985: 5). This resulted partly from technological change and the upturn in the economy, but was also due to the retention of classified advertising and the continued erosion of the HWT's share of display advertising (*Annual Report*, 1985: 5). In 1984/5, Syme achieved market dominance in display advertising as well as classified advertising for the first time, with *The Age* carrying 71.4 per cent of Melbourne classified advertising, the market shares of display advertising being 36 per cent to *The Age*, 34.7 per cent to *The Herald* and 29.3 per cent to *The Sun* (*Annual Report*, 1985: 29, 30).

The structuring of daily newspaper publishing in Melbourne creates pressures on proprietors to continue publication to retain both readership and advertising. This has generated a tendency on the part of proprietors to maintain consensual industrial relations policies and collective bargaining and negotiated agreements with employees to reduce uncertainty in the market. With the volume of classified advertising carried by *The Age* and the size of the newspaper, these pressures were intensified in the transition from 'hot metal' to 'cold type'.

The negotiated agreement

Technological change is an uneven process. Different industries and organisations exhibit different patterns and rates of technological change which reflect different product and labour markets, worker (and managerial) resistance and organisational cultures. As indicated earlier, technical change is not new to newspaper printing. The introduction of the Linotype machine represents the last major instance in typesetting. Fitzgerald reports that the printers hoped David Syme would not introduce the Linotypes because of his highly profitable operation (1967: 76–7). As a capitalist enterprise, the Syme organisation could only delay the adoption of the new technology (Hagan, 1966: 104). With the introduction of the Linotype in 1897, *The Age* 'companionship' (as printing craftsmen call themselves) was reduced from 120 to 44 (Hagan, 1966: 104) from which the final choice of operators was made. Consistent with his liberal–humanist values, but most importantly to maintain the worker loyalty on which his industrial relations practices relied (Lawson, 1981: 55), Syme arranged pensions for older employees and established others on farms and in newsagencies (Pratt, 1908: 265; Sayers, 1965: 221, 303).

There are both continuities and discontinuities in the change to computerised photocomposition in the 1980s. Computerised photocomposition came later to Melbourne than to the United States or even New South Wales newspapers (Zimbalist, 1979; Souter, 1981: 559–67). New Linotype machines were no longer available and the managerial strategy of publication based on a near monopoly of classified advertising could not be sustained without the capacity to meet market requirements. Further, the extent of classified advertising at *The Age* prevented early adoption of the new technology since a system capable of handling the volume had yet to be developed. The organisation claims to carry the largest volume of classified advertising in the world in its Saturday edition, which currently runs to 238 pages. Once such a system was developed there were significant market as well as labour cost-saving pressures on the Syme organisation to adopt.

As noted above, computerised photocomposition was introduced into the production process at *The Age* between 1980 and 1983 on the basis of negotiated agreements concluded in 1979 between the Melbourne Metropolitan Newspaper Publishers' Association (MPA) and the three unions claiming at least some jurisdiction over inputs of copy to the system (the PKIU, VPOU and AJA). The key agreement is that concluded between the MPA and the PKIU, signed on 10 July 1979, Although print workers initially claimed control over entry of all copy to the computer, agreement was based on a ruling by Justice Cahill in the New South Wales Industrial Commission in 1977 (No. 114 of 1977) following industrial action by members of the PKIU employed by the Fairfax organisation in Sydney concerning the conditions under which the PKIU would accept the introduction of computerised typesetting (Souter, 1981: 560–7). The Cahill judgment was formulated on the principle of the retention of the first keystroke or 'direct entry' at the point of origination of copy within the organisation.

The allocation of tasks within the technical division of labour outlined in the Melbourne agreement provided for the direct input of editorial copy of employed journalists and of telephoned classified advertisements received by employed telephonist–clerks. Former hand and machine compositors (printers) retained control over 'hard copy', that is, contributed editorial material, letters to the editor, etc., and classified advertisements received by mail and over the counter. Two subsequent disputes between the PKIU and the Fairfax organisation concerning 'tipping panels' in horse-racing, trotting and greyhound-racing and the input of television programs were resolved. In the first instance, the PKIU claimed jurisdiction over what had been AJA occupational territory. In the second instance, the AJA and the PKIU, both supported by the Labor Council of New South Wales, contended that this was PKIU work. In both instances the commission found for the proprietors, that is, the Fairfax organisation (No. 22 of 1980 and No. 339 of 1981), and both areas remain under the jurisdiction of the AJA. In Victoria, clerical workers who were formerly members of the VPOU and who compiled television programs prior to computerisation are now low status members of the AJA. A further challenge to this division of labour on the part of the PKIU, supported by the AJA, seeking PKIU jurisdiction over telephoned classified advertising copy, resulted in indefinite adjournment by Deputy Commissioner Keogh in June 1984 on the grounds that it would not be 'in the public interest'.

The technology agreements provided for retraining for full-time employees and limited health and safety provisions, for example, a break of ten minutes for VDT operators during each half of a full shift, eye tests 'as appropriate' and consultation with the union on lighting and seating arrangements.

Printers were the occupational group for whom the change in technology and associated job redesign represented the greatest threat to employment and change in work organisation. Guarantees of 'lifetime' employment were given for permanent employees as at 31 October 1978, together with provision for compensation for earnings forgone by former full-time Linotype piece work operators and provisions for voluntary redundancy on an age basis with compensation for lost earnings. Where redundancies were to be declared, provision was made for 'voluntary termination of employment', with preference for employees aged 60 years or more. Redundancy provisions included severance payments at the rate of a minimum of three weeks' pay for each year of service, extended on a pro rata basis for each additional completed quarter of service. No overtime, shift allowances, penalties or margins were included and payments were not to exceed anticipated wage payments had the worker remained in employment until retirement. Those with seven or more years' service were to receive pro rata long service leave.

Furthermore, in keeping with the provision concerning 'retention of wage rates', compensation in the form of three instalments over a period of two years was to be made to *full-time* piece work Linotype operators when transferred to other duties. This payment was based on an updating of a 1967 agreement designed originally to compensate workers who moved from Linotype piece work to TTS operation. At the same time, and in line with the PKIU preference for protection of jobs rather than 'golden handshakes' to leave the industry, other concessions were negotiated, such as progressive reductions in hours to a 37 hour week (36 hours night work), an additional three days' annual leave and increased time for authorised chapel meetings and trade union training.

The outcome—gender and skill

As a result of the technology agreements, 60 printers were made redundant, some investing redundancy payments in small businesses and hobby farms. Those remaining were retrained, both through outside typing tuition and 'in house' system training, to use the electronic keyboards and VDTs, Xenotron displayers or page paste-up. It is possible to see these former tradesmen as deskilled. Highly valued skills like hyphenation and justification to provide appropriate word breaks and line lengths are no longer required. The ability to conceptualise and assemble pieces of metal and lines of type into a newspaper page with only reverse images of letters for guidance has become obsolete with 'cut and paste'. The work they now do, whether keyboarding copy, displaying advertisements or 'page paste-up' is highly specialised. The keystrokes of the workers are not counted, although workers are aware of the potential. As former printing craftsmen, however, they bring to their work knowledge of printing, newspaper style and presentation. A residual craft pride in quality production informs their work alongside a sometimes angrily, sometimes sadly, expressed sense of loss of interest in 'standards' on the part of management, journalists and other production workers.

The chapel at *The Age* continues to represent these workers' material interests to management. Chapel representatives continue to allocate overtime and to prepare the rosters for holidays and long-service leave. Although the potential to monitor keystroke rates exists and printers are aware of it, agreement not to do so has been negotiated, due in part to the incidence of repetition strain injury. These tradesmen, as is consistent with their occupational identity, continue to exercise controls over time, despite the deskilling and management control potential of the technology of computerised photocomposition. The PKIU, although significantly reduced in membership as a result of technological developments, continues to pursue its members' interests, particularly through further union amalgamations within the industry (Bennett, 1979: 35–6).

As Cockburn (1983) has shown, apart from the issue of control, much of the opposition (and not a little pain) associated with computerised typesetting is linked to notions of masculinity and 'men's work' and the association of notions of femininity and 'women's work' with the QWERTY keyboard of the new technology. Women are assumed to be 'naturally' good typists, therefore the increased skill in keyboarding for direct input on the part of telephonist–clerks receiving classified advertising is not recognised. Women bring typing skills with them to the job rather than acquiring them on the job (as in apprenticeship). Similarly, the inter-personal skills used in effectively dealing with the public through the medium of telephone are not regarded as skills. Receiving and canvassing for advertising represent the exploitation not only of women's mental and manual labour, but also of their emotional labour (Finch and Groves, 1983: 1–10). On the basis of the analysis of the redesigned jobs, it appears reasonable to see the jobs of the former craftsmen printers as deskilled—even 'de-masculinised'. The routine female telephonist–clerks' jobs are more appropriately seen as 'upgraded'—an 'upgrading' rarely recognised and not rewarded through increased remuneration (see Reed, 1987). For journalists the outcome is uneven, with reporters experiencing little change and sub-editors carrying increased responsibility as well as control over their product. These, however, are issues outside the scope of this chapter.

The outcome—job satisfaction

Job satisfaction research is an enterprise fraught with pitfalls (Kahn, 1972; Locke, 1976; Emery and Phillips, 1976). Lowered expectations (Salaman, 1979: 155), the position of women within the domestic division of labour, the ideological effects of the 'work ethic' (Anthony, 1977: 22), and the problems inherent in the identity-conferring nature of work in advanced capitalist society combine with other extraneous factors to make the measurement of job satisfaction a notoriously unreliable task.

Research has, however, indicated consistently that work satisfaction for craft and professional groups is most influenced by intrinsic factors (Voydanoff, 1978: 190). On the basis of the tradition of print as a male craft occupation, former printers were asked, 'Would you mind if your son decided to do your type of job?' and 'Why?'. Other measures which reflected the occupational cultures of telephonist–clerks and journalists were also developed. The results indicated that both telephonist–clerks and reporters were satisfied with their redesigned jobs. Sub-editors showed ambivalence concerning the tensions between increased control and increased responsibility in their new jobs. Printers' responses indicated dissatisfaction reflecting the deskilling and de-masculinising aspects of the new technology. A dissatisfied printer summed up: 'The work used to be skilful. Now skilled labour—not tradesmen's work. He should choose a different trade—or journalism.' Another said, 'It would be OK for a daughter. Not demanding enough for a man as it is now.'

Conclusion

There is a striking similarity across societies in the extent to which printers perceive their jobs as de-masculinised because of the identification of 'hot metal' production with 'men's work' (cf. Cockburn, 1983; Hill and Novitz, 1985). In no instance, however, after the initial United States confrontation (Zimbalist, 1979), was the new technology rejected (cf. Hill, 1984: 167). Struggle has centred on control over the entry of copy to the computer and the distribution of rewards from increased productivity. Australian agreements centred on the capture of the first keystroke or direct entry of copy to the computer.

Initially, United Kingdom and New Zealand agreements centred on re-keyboarding to avoid redundancy (Hill, 1984: 174; Cockburn, 1983: 71, 73, 79, 83). In the longer term, New Zealand printers have accepted direct entry without redundancy on the part of tradesmen, but at the expense of female non-journeymen (sic) TTS operators. United Kingdom newspaper printers' experience is mixed with some redundancies (Cockburn, 1983) and continues to be characterised by conflict between and within the ranks of proprietors and unions alike (Martin, 1981; Cockburn, 1983). With the collapse of the bitter year-long dispute surrounding News International operations at Wapping (*Sydney Morning Herald*, 7 February 1987), it appears that resistance based on strike activity is likely to result in non-union shops as in the United States or, as in the United Kingdom, electrical union control.

The Melbourne agreement is regarded as an achievement by the PKIU and a good example of its type in industrial relations terms (Riches, 1980: 203; Deery, 1982: 171–2). It is, however, a 'defensive' agreement to facilitate technological change, although on terms which limit management's power to rationalise labour and exclude union influence (Riches, 1980; Deery, 1982; Manwaring, 1982; Gill, 1985: 124).

Unlike the New Zealand Printing Union, the PKIU has bargained consistently in ways which excluded female workers from production. In doing so, it drew on 'craft' values to legitimate its claims—a value rationality based on the notion that 'letter assembly' is the work of compositors. Consistent adherence to this position on the part of the union avoided the necessity encountered by New Zealand printers of arguing against their previous claims (i.e. that TTS was *not* tradesmen's work), for control over the new technology. The existence of the VPOU was an obstacle to PKIU control and advantageous to proprietors seeking to reduce costs, but it also protected the interests of female telephonist–clerks in Melbourne. New Zealand printers have avoided redundancy among compositors (Hill, 1984; Hill and Novitz, 1985) through maintenance of male interests over those of female TTS and 'tele-ad' operators who have non-journeyman (sic) status in the Printing Union (Hill and Novitz, 1985: 18), but it remains an open question, as in Australia, whether the strength of the union can be sustained in the long term (Hill and Novitz, 1985: 39)

Printers in Australia and New Zealand were also aware of developments in the United States, the source of the technology. For a country like Australia (or New Zealand), with its intermediate position within international capitalist relations, imported technologies, developed under specific manifestations of capitalist class relations in metropolitan countries like the United States, carry those relations into organisations employing that technology. Competitive pressures in the United States newspaper market, which coincided with a veto of technological change held by the International Typographical Union (ITU) (Zimbalist, 1979), intensified investment in a search for a technology which would break the power of the ITU (Smith, 1980: 77–9). The extent to which alternative task allocations were likely with the imported hardware *and software* appears limited given the goals of management to increase the volume of advertising carried with closing times as late as possible.

This study demonstrates the importance of extending the labour process approach through Weberian concepts like bureaucracy (e.g. Littler, 1982), closure theory (Parkin, 1979; Murphy, 1986) and rationality as an analytical construct (Brubaker, 1984) to provide a more adequate explanation and interpretation of instances of technological change. On the basis of the above, the redesigned jobs of Melbourne newspaper printers appear to be deskilled. At the same time, it is indicative of the historical strength of the union and its substantively rational strategies in industrial conflict that tradesmen printers are still employed in the production process at *The Age*.

4 New technology outwork

JUDY WAJCMAN AND
BELINDA PROBERT

The case studies in this book are all concerned with the organisation of work in particular industries and workplaces, and they are informed, albeit critically, by what has been called the labour process tradition. Empirical research such as this has helped to dispel oversimplified conceptions of workplace relations, derived from general theories of managerial omnipotence, or of social consensus, for example. It is often difficult, however, for labour process research to reflect seriously on more general trends in the organisation of work which are determined by an increasingly global political and economic framework of capital accumulation. Yet this broader perspective has become increasingly necessary with the onset, in the early 1970s, of a profound economic recession affecting all the advanced capitalist economies. Associated with this recession we have seen massive international shifts in the location of work, major restructuring of many industries around new technologies, and increasing conflict over existing employment practices, focused on demands for the deregulation of the labour market.

An important feature of this prolonged recession has been the loss of full-time permanent jobs in the OECD countries. One indicator of this is the increase in officially recorded unemployment levels, which in Britain, for example, have risen to 12 per cent (OECD, September 1986), and in Australia to 8 per cent (ABS Labour Force Statistics, January 1987). Equally significant, but generally less well publicised, has been the rapid increase in part-time employment. In Australia, over half the new jobs created since 1973 have been part-time. Alongside these quantifiable changes there has also been an increase in employment outside the regulated sphere that is difficult to measure. For example, outwork that is paid for cash-in-hand has increased, while much part-time work lies outside the reach of protective industrial legislation. Changes of this kind must be placed in the context of the generalised capitalist recession and the changing global division of labour, rather than the condition of local labour markets.

There are other important changes occurring in the organisation of work which also suggest a trend towards deregulated employment. There appears to be worldwide increase in the recruitment of workers on short-term contracts, in the use of sub-contractors, and in self-employment. In the Australian retailing industry, for example, large firms cut costs by dealing with suppliers who are sub-contractors (middlemen) and who in turn may employ outworkers (on rates of pay that are partly determined by competition from Korean labour markets). Such sub-contracted employment relationships are extremely hard to regulate since there is no direct link between the employer and the outworkers (employees). This form of sub-contracting is not confined to traditional industries with old-fashioned technologies such as the clothing industry with its sewing machines. Sub-contracting and self-employment are also common in some new and flourishing high-technology occupations, such as computer programming.

In Australia the increase in self-employment has been substantial, rising from 427 800 in 1972 to 746 000 in 1986 (ABS Labour Statistics, August 1986). Most of this increase can be accounted for by increases in sub-contracting and in various kinds of outwork, employment situations often associated with inferior working conditions. It is important to recognise, nonetheless, that the self-employed are not necessarily disadvantaged *vis-à-vis* employees, and that their income and status will depend critically on the strength of their position in the labour market and, for example, their ability to exploit the tax system to their advantage.

Despite the scale of the changes in work organisation that we have just described, the labour process literature of the last fifteen years is still overwhelmingly preoccupied with traditional male unionised sections of manufacturing industry. For example, discussions about the impact of new technologies on the workplace have focused to a remarkable degree on the automation of machine tools (Noble, 1984; Blackburn et al., 1985; Shaiken, 1984). Yet at the same time the proportion of the workforce employed in manufacturing has steadily declined. In Australia, only 16 per cent of the labour force was employed in manufacturing in 1986 (ABS, August 1986). As well as having a rather narrow industrial focus, most labour process work has also confined itself to the tendency of new technologies to deskill workers. For all these reasons, our study of new technology outworkers has moved us on to less well charted terrain. We are looking at people who work with computers, both those whose job it is to program them and those who process information on them. In particular, we have looked at those who do such work from their homes, most of whom are in some sense self-employed. In contrast to studies of male, unionised manufacturing workers, we have studied a group of men and women who possess new skills that are required by the new technologies, who are non-unionised, and whose workplace is the antithesis of the industrial factory.

Until recently working from home, or outwork, had been considered an anachronism, largely confined to the traditional and declining manufacturing sectors. In Australia, research has focused on, for example, the concentration of migrant women in unregulated, non-unionised employment conditions in clothing and textiles. However, since the mid-1960s outwork has been growing in information processing and the high-technology manufacturing sectors, as well as expanding again in traditional industries (Lipsig-Mummé, 1983). Increases in traditional forms of outwork, often characterised as 'sweated trades', can largely be accounted for by the competitive pressures emanating from a new global division of labour, the threat of offshore production and a continuing supply of marginal workers such as migrant women.

Research into new forms of high-technology outwork typically approaches the field with a set of assumptions derived from the study of traditional outworkers. Huws (1984), for example, who studied female computer programmers, expected new technology outwork to become more widespread because it is a method of production favoured by employers seeking to resist competition and protect profits. Outwork can reduce employers' costs in a number of ways: wage costs are reduced by isolating workers and preventing unionisation, on-costs and fixed-costs such as award entitlements are avoided, and productivity can be increased by changing the payment system to one of payment by results. It is this payment system which also provides the chief mechanism by which the output of homeworkers who are not directly supervised can be controlled (Allen and Wolkowitz, 1986: 29–31). In the case of computer-based outwork, it is assumed that, in addition to the simple control exerted by such a payment system, the computer itself can enhance employer control, for example, by monitoring the key strokes of homeworkers.

In stark contrast to this somewhat pessimistic approach to new technology outwork there is a more speculative (and widely publicised) literature which celebrates the dramatic changes to work and family life that are now possible. It is argued that recent technological innovations permit small, decentralised production units to be an efficient alternative to large centralised organisations, and that 'flexible specialisation' in product strategy requires analogous flexibility within the workforce. The suggestion is that new technology serves to increase rather than reduce skills and that future production will be 'high-technology cottage industry' (Sabel, 1982: 220). Alvin Toffler is probably the best known exponent of the view that such changes will lead to the reconstitution of the home as a workplace, with the nuclear family as a new work collective. 'The rise of the prosumer, the spread of the electronic cottage, the invention of new organisational structures in business, the automation and demassification of production, all point to the home's re-emergence as a central unit in the society of tomorrow.' (1980: 364). Within this home, an egalitarian sexual division of labour is now possible, based on the reintegration of work and child care. In this idealised picture of the home life of the future, it is the potential of new technology to serve individual interests that explains the growth of outwork, rather than the competitive pressures of an increasingly global capitalist economy.

Despite much speculation about a future in which millions of people work at home, linked to central offices with computers and telecommunications technology, there has been no research in the area in Australia, and very little anywhere else (Huws, 1984: Olson and Primps, 1984; Pratt, 1984; Vedel, 1985).

The study

Our study is based on interviews carried out with two groups of new technology homeworkers. The first goup, computer programmers, was chosen because it appeared well placed to exploit the advantages of homework so glowingly described by Toffler. The second group, word processors, was chosen because such workers seem, on the surface, vulnerable to many of the pressures faced by traditional manufacturing outworkers.

Our sample was drawn from New South Wales, with the exception of two programmers from other states. In all, 39 homeworkers were interviewed, of whom two-thirds (26) were programmers and one-third (13) word processors.

The programmers were found by advertising in newspapers and Telecom's Viatel system, through word-of-mouth and software company contacts, and through computer user groups. The word processors were found by following up advertisements in newspapers and on notice boards, and by advertising through community-based centres and word-of-mouth. The interviews were mostly (28) conducted face to face, with three being done by telephone and eight by post.

Approximately half the sample (20) were women and half (19) men; most of the programmers were men (18) and the word processors, with one exception, were women (12). Although their ages ranged between 21 and 55, about two-thirds were in their thirties, and the great majority were married or in de facto relationships (87 per cent). The sample was a very highly educated one, with 85 per cent of programmers and 46 per cent of word processors having tertiary qualifications, and 82 per cent of them were either Australian born or from other English-speaking backgrounds. In these last two respects our sample differs sharply from traditional outworkers, 85 per cent of whom are estimated to be from non-English-speaking backgrounds, and only 17 per cent to have post-secondary education (Hendricks, 1986).

Most of the word processors had been working at home for less than two years, while the bulk of the programmers had been doing it for more than two years (though no one had been working at home for more than twelve years). The most common reasons they gave for working at home were the flexibility and convenience of such work patterns, a desire to be close to their children, a preference for being one's own boss, and in order to reduce overheads. As 'flexibility' was a major reason for homework, we asked people what they meant by the term. Half of the sample said it meant choosing your own hours of work, and many others mentioned independence (5), setting your own priorities (6) and being able to choose jobs and clients (4). Other factors mentioned were the ability to combine work and child care (5) and the ability to adapt to clients' needs (9). Cindy, a word processor, gave these reasons for working at home:

> I have flexibility to work the hours I want. I can theoretically say no to a job if I don't want to do it . . . I can go off and do an aerobics class in the middle of the day or something, or go for a walk.

Alex, a programmer, said in similar vein:

> I want flexibility because I have found I enjoy being my own boss. I had six months where I thought I was developing an ulcer. But it [homework] actually works and I have a nice sort of lifestyle.

It is quite clear that many of our homeworkers were strongly motivated by a desire to be their 'own boss'. It is also striking that for both programmers and work processors there was a genuine element of choice involved. In both these respects their experience is quite unlike that of traditional outworkers.

There was also a real sense of choice in the replies of those who cited a desire to look after their own children as a major reason for working at home. About half of the sample (24) had children under twelve, and thirteen of them said that they wanted to be close to their children. It was, however, overwhelmingly women who gave this as their main reason for doing homework, with only two men mentioning it. As Brenda, a word processor, said:

> No, I preferred not to go to child care. I believe a two-year-old should be with her mum, even though now I think that one day a week would be fine, get her out of my hair. I didn't even look at child care . . . and also getting into the scenario of telling my hubby how many times I had changed the baby's bum, and I thought I am not going to get into that. I'll keep myself interested and keep myself within the industry, keep the brain ticking.

For many of these women, however, it is a choice constrained by a lack of alternatives. Cindy gave the following account of how she experienced this:

> even though my children are, I think, at a reasonable age now, they are just not reliable kids. One day I came home from work having left the children here after that school period because they didn't want to go to after-school period, because that was for babies, and I found they were making cross-bows using nails. There was a lot of blood in the sink; all the neighbourhood children were here and they all had cross bows. And I thought this is the sort of thing that needs parent supervision, just slightly. So I decided that that was it.

Another word processor, Olive, remarked:

> part of the idea of working at home is because I don't want to send the children out to care. If there was good child-care available in the work-place with my access to it—I mean physical access to it, I would change my attitude, although with a six-month-old baby I still don't think I would. With a two-year-old it is better.

But for some women, what starts out as a temporary solution to the problem of combining paid work with the care of very young children becomes a trap. Sue, a programmer, experienced this:

> I wanted to work part-time because I have got a son who is sixteen months old, and I didn't want to miss out on him growing up because he is changing so much at the monent that I thought I would miss a great deal by having someone else minding him all

> the time. And the place I was working in, when I was pregnant and had maternity leave, I hoped that I could go back part-time, say two to three days a week. Anyway, that didn't work out at all and they said wait seven to eight months and we might have something.

Regardless of the type of work they do, women with young children are in a weak position in the labour market. It is because of this that they are over-represented in the categories of part-time work and outwork. Motherhood leads to labour market vulnerability for several related reasons, the most important of which is women's primary responsibiltiy for child care within the family. This socially defined sexual division of labour within the family is reinforced materially by an acute shortage of the kind of affordable and good quality child care which would be necessary to permit full-time employment with a normal working week. And the ideological cement is provided by women's own desire to take care of their children when they are very young.

In the idealised picture of the future presented by enthusiasts for the 'electronic cottage', this domestic dilemma is resolved when men bring their work back into the home, and become available for greater child care responsibilities. However, only one man in our sample gave child care as a major reason for working at home, and he was a sole parent. There is an important difference, of course, between being at home and being available for child care. When we looked at the actual working habits of our sample it became clear that their portrayal as harbingers of a more egalitarian and child-centred way of life is, indeed, idealised. For despite their references to having choice in and control over their manner of working, our sample in fact worked very hard and for very long hours. Only two people worked a usual working week (35–40 hours). Word processors tended to work less, and programmers tended to work more, at night as well as during the day. All of the programmers worked at weekends, as did many word processors. In fact two thirds of the sample said they did not have a day in the week on which they did not work, and there was little difference between programmers and word processors in this respect. Given work patterns such as these it seems unlikely that the traditional domestic division of labour is being radically transformed, nor is there any evidence of the increased leisure and fuller, more creative lifestyles that the optimists predict.

Not only did our sample tend to work very long hours, but two-thirds of them said they worked more productively at home; 81 per cent of programmers and 36 per cent of word processors attributed their increased productivity to the lack of distractions at home. A more comfortable work environment and choice of working hours were the next most frequently cited explanations. Mary, a programmer, said:

> I prefer to be on my own. The reason I work from home is that I get the peace and quiet—I don't get the distractions. I don't like having people around me. That is the thing that annoys me when I am at the office . . . because you are in the middle of some train of thought that has taken you ten minutes to get there, the phone rings or somebody walks in just to say hello or chat about lunch or something silly, and you have lost that train of thought and it takes fifteen minutes to get back to where you were. But then I am the kind of person who likes to work on my own anyway . . .
>
> I get twice as much done at home and in a much more relaxed fashion . . . I am a good worker in the morning. I do my most creative stuff early and if I get up at six in the morning and I want to start at 6.30, two minutes out of bed and I am at work, you know. On the way I make my first cup of coffee and I can be working and then when you're compiling something or doing something of that sort for five or ten minutes that you cannot physically do anything, I go out and I put the washing on, or I go up the street and I get some food. Or, you know, 4 p.m., if I am tired I will take a break and I'll go and sit in the sun for a while and then I will come back.

The presence of children and other distractions were cited as causes by the 27 per cent who thought they were *less* productive at home. As Olive pointed out:

> There's always interruptions. It's very hard to get a clear hour without some form of interruption because of the telephone, or personal life or babies ha! ha! . . . My real work is running this business of a house and I need to get away from it to get other things done.

One word processor, Vicky, described very vividly the combination of advantages and disadvantages that a homeworker with children might face, when we asked her whether she was more productive at home:

> No, given the constraints I have. That is a very tricky one, because left to my own devices I am. For example, I had to do something fairly urgently on the weekend . . . for eight hours which I think was fairly productive. And I made the deadline—I don't think I could have done that in an office environment, there are too many interruptions—because I made specific arrangements for this . . . I got rid of the kids . . . [And the constraints?] First of all the time and the fact that I generally leave it until the baby is asleep; that it is fairly impossible to try to do work without

> interruption with a two-year-old around so that's the largest constraint . . .

As one of Huws' respondents put it (not without a sense of irony), the main advantage of working at home is 'being with children all day' and the main disadvantage is . . . 'being with the children all day' (1984: 43).

Thus far we have been drawing attention to the way in which the traditional sexual division of labour continues to shape the work patterns of the homeworkers we interviewed. But it is also the case that the different nature of the work done by programmers and word processors affects their respective patterns of work, particularly the regularity with which it takes them away from home. In fact, the major external constraint on getting on with work at home cited by our sample was the frequency with which they had to travel to work sites or clients.

Visits to client sites by programmers varied in duration and frequency, depending on the stage of the work and whether or not problems could be solved over the phone. At the installation stage (when the software was being installed into new machinery at the client's workplace) the programmer was on site full-time until the client became familiar with the system. Subsequent visits for maintenance to that particular client then became much less frequent. Overall, 62 per cent of programmers regularly visited sites two or more times a week. By contrast, 63 per cent of word processors visited a client or work site once a week or less and their excursions more closely resemble those of clothing outworkers.

Most of this variation can be explained with reference to the precise nature of the work carried out by the two groups we studied. After initial discussions with the client, software 'design' (mapping out the type of program required for the client's work) can be carried out in the programmer's office or home. In some cases, depending on machine and language compatibility, the programmer can at least write draft programs in an office or home, to be installed afterwards. This potential has been greatly enhanced by the advent of the personal computer. After installation, however, the work of a programmer involves regular visits to the client's site so that the software 'bugs' can be removed, and modifications made. In addition to these specific site visits, it was common for our programmers to be 'attached' to a software house, and to make regular visits to the base office.

The work processor has a very different pattern of work. Most of those we interviewed found work by advertising in local and student newspapers, and their clients tended to come from their own suburb or close neighbourhood. While programmers visited their clients on site, word processors received theirs at home. The occasional visit that the word processor would make was either because they provided a pickup and delivery service with their work, or because a client required their word processing skills (and not their equipment) on site. In this latter case visits to a client's site resemble a form of 'temp' service, but this was by no means common. Outwork is usually thought of as being work done in the home, with occasional trips out, for example, to deliver completed work. Most of the work done by our word processors fits this model and, like traditional outworkers, many of them feel isolated as a result.

The programmers in our sample, on the other hand, did work which fits neither the model of full-time outwork nor full-time site work. And it is precisely the fact that they can combine working from home with working on site that they like. It is the programmers' privilege to be able to enjoy the best aspects of each. So while Toffler idealises a totally home-based working pattern, the evidence from our study suggests that it entails significant social costs.

Programmers appreciated their flexible and varied working patterns but, as we pointed out earlier, the very long hours they worked militated against any significant change to the balance between work and leisure, or between work and family life. When we asked both programmers and word processors if and how working from home had changed their attitude to work, we found strong evidence of reinforced rather than transformed gender differences. Most (84 per cent) of the sample said their attitude to work had changed, with over half (57 per cent) saying they were more work-centred. But this was especially so for the men. Around a quarter (28 per cent) of the sample, all of whom were women, said they were less work-centred and more family-centred. Similarly, increased gender differences are reflected in the fact that nearly half (46 per cent) of the programmers thought they had become *more* workaholic, while most (83 per cent) of the word processors were *less* workaholic. David, a programmer, talked about this problem:

> I would say quite honestly that the prime consideration in my life is my work . . . and that has caused all sorts of problems, personal problems and I am the first to admit that I don't spend enough time doing the normal things that people do. I tend to think, live and breathe my work, umm [notices something come up on computer screen] . . . I mean, to stay away from a computer for any length of time is like withdrawing from heroin.

His wife commented:

> I am not kidding, you can be having one of the most important discussions with him and if you are anywhere within visual sight of

> that thing he will be keeping one eye on it the whole time . . . we can be talking about a life and death situation and [he] has got one eye on the screen.

The contrast between the attitudes of the male programmers and those of some of the women word processors could not be more stark. As Brenda put it:

> Yes, for me work has lost a lot of its importance . . . I mean lifestyle . . . for example most men don't know their children . . . you have got to know your children, it's very very important . . . lifestyle and the family unit . . .

And Olive said: 'Over the past six years it [work] has assumed less [importance]. It was very high when I started. Not working, children have replaced it.'

From what our programmers have said it is clear that the creative aspect of their work is precisely the aspect that makes them obsessive. Yet this is an aspect about which the labour process tradition has little to say in as much as it has tended to focus on the deskilling effects of new technology. In particular, labour process analysis has emphasised the increasing separation of conception from execution at the workplace, and the use of management's monopoly over knowledge to control the labour process. Software programmers, however, are prototypical knowledge-workers engaged in genuinely intellectual work, which partly explains the way they organise their work. The autonomy and creativity involved in programming make it the kind of work that accords well with optimistic scenarios about the future of work in the home, such as Toffler's. It is ironic that it is precisely these qualities which in fact lead to obsessive work patterns in the home, where there are fewer constraints.

It is important to note that women word processors also sometimes work extremely hard, at all times of the day or night. But this is not a result of the nature of the work itself. Rather it reflects the erratic pattern of demands for their services, externally imposed short deadlines, and their individual tolerance of unreasonable demands. Cindy frequently feels unable to refuse work, despite the problems it can generate:

> It does get out of hand and that is one of the difficulties. I am afraid I am a bit soft and I get talked into things. I feel sorry for young kids who have obviously done work . . . especially in schools, and it's got to be there at 10 a.m. tomorrow morning. So I tend to work very late that night and get it done for them and get very tired and feel absolutely exhausted the next day and the family suffers too because they don't get a proper meal cooked for them and I am not looking at the kids' homework that I should be attending to . . .

Justification for working at local centre, not home

From bitter experience Brenda has learnt to be ruthless with clients:

> I had this idiot ring me up last night and I said to him, well, couldn't this have waited until tomorrow because there is nothing dire straits? I said well, look, I don't mean to be rude but didn't your mother ever teach you about, you know, you don't ring people up after 9 p.m.? I have just got to the point where I say, look if you want to be like that I don't work for you . . . I plan my day, say three to four people stretched through the day, which gives me an hour, hour and a half break in between work. You make sure you plan it that way. If someone says it's 5 p.m., can I come around at six, its obviously inconvenient because you have got a dinner to prepare, sit down and have a fag and be a zombie for an hour.

The way in which people work is also clearly related to the way in which they are paid, and here again there are significant differences between programmers and word processors. Almost all the word processors (91 per cent) were paid piece rates, hourly rates or a combination of both. The most common payment system for programmers (88 per cent) is a combination of fixed price and hourly rates. The piece rate system means that the worker is paid by results, that is, for items or words actually produced. Under this system the pay is the same regardless of any unforeseen difficulties that the worker may encounter. It is a payment system which serves as a form of direct control over the worker, and one which is responsible for low pay (Allen and Wolkowitz, 1986).

Much has been written about the extremely low piece rate payments made to traditional outworkers. Although our word processors earned much more, it is important to note that 77 per cent of them said they were earning less since they started working at home. A quarter of our sample said they were charging less than market rates and many reported pegging their rates according to their impressions of what other women doing similar work were charging. As Cindy said:

> I have had a few phone calls that I suspect are from people sounding me out a bit, ringing up fairly obviously asking about prices. I don't know if they are my competition or what.

Another word processor, Ann, ventured:

> It's difficult, but I try to check what the industry is charging people working in the same field . . . [how do you do that?] I just ask around what people are paying . . . I invariably find that I am lagging behind . . .

Employee Issues
— How paid?

Interestingly, word processors did in fact vary their piece rates, at some cost to themselves, by generally charging students less ($1.50 per page) than their more affluent clients ($2).

The hourly rate does at least give word processors some pricing flexibility, reducing the pressure for speed and allowing for items that are particularly difficult to type. This rate was applied to proofreading, alterations or re-drafts and consultation regarding format and layout. But, once again, prices varied according to the client's ability to pay. The concession rate was about $10 per hour, while the full rate was about $12 per hour, this being the hourly rate for 'temp' work. This compares with the award rate for clothing machinists of $6.22 per hour, while clothing outworkers would often earn much less again.

Working at home had changed the earning patterns of the great majority of our sample, but the nature of the change varied according to the type of work. Most of the word processors (77 per cent) were earning less, whereas half of the programmers had increased their income. For many of the latter, however, their earnings were very uneven. As Don said:

> Yes, some weeks I will bring in $1400 and other weeks I will bring in $600 but it evens out into a fairly reasonable income in the end, and working as a sort of a one-man company I have got enough tax perks still to make it worthwhile.

Hardly anyone in our sample, however, took into account the real costs of working from home, such as electricity, equipment maintenance and hire, phone calls, travel, training costs and so on.

The important difference in the payment systems of programmers and word processors is that the former are not paid by piece rates. Most programmers preferred the hourly rate over the 'fixed-price' system as it allowed for unforeseen difficulties. The hourly rate charged varied according to the client's ability to pay, and the duration of the project, with short-term projects attracting higher rates than long-term ones. The variation in these programming rates was considerable, with some as low as $15 to $20 per hour and some as high as $50 with the most common being $30 to $35. Most of the programmers we surveyed avoided the fixed-price system which operates extensively within the computer industry. Under this system a price for the entire job is quoted, based on an hourly rate and estimate of the project's duration. Many programmers said they had been 'caught short' because they could not accurately predict the completion of the project. Only the very experienced contractor would venture into this system after making one such mistake.

Despite the difference in rates of pay, the vast majority of

employment status

programmers and word processors shared the same employment status, being either 'self-employed independent contractors' (10 and 9 of our sample respectively) or 'employees of own company' (14 and 2). The distinction between these is that the 'employees of own company' have formed their own companies, incorporated or family, and these stand between them and their clients, giving them company taxation advantages. In many cases, working from home was a way of reducing the overheads of a fledgling business.

New technology homeworkers who are self-employed contractors share this employment status with many traditional outworkers. The growth of this kind of self-employment, which is not covered by industrial legislation—in other words, unregulated employment—has been viewed by many with great concern. It involves the loss of the considerable benefits that attach to employee status. The word processors in our sample, for example, were not insured against sickness and were not covered by any holiday or long-service leave provisions. Only two of them were covered for injury (worker's compensation), and only one contributed to a superannuation scheme. The programmers looked after themselves rather better, 70 per cent of them being covered for injury, and just over half having insurance against sickness. They had no holiday or long-service leave provisions, however, and only 35 per cent of them contributed to a superannuation scheme (though it should be noted that in Victoria outworkers are now covered against injury under Work-care). However, most other studies of new technology outworkers have focused on those with employee status, and they suggest that in fact this status does little to protect the employment conditions of those who move home to work (Huws, 1984).

A major difference between being an employee and being self-employed is, of course, that the latter must pay for their own equipment. In the case of computers this equipment is very expensive. All the homeworkers we interviewed had one or more micro-computers, except one who relied on a link to a mainframe computer, and all of them had one or more printers. A quarter of the sample had ergonomic chairs, and half had answering machines. Of the programmers, thirteen had a modem and three had a mainframe link. About half of the sample had purchased their own equipment, and the rest were either leasing it, or had it on loan, and five women had had it bought for them by their spouse. The programmers found it hard to estimate the cost of their equipment, and the sums varied widely. But even in the case of the word processors' equipment, over half the sample paid between $10 000 and $15 000.

In addition to this equipment, most of these homeworkers, programmers and word processors alike, said they had an office. Not surprisingly, then, their houses suggested a fairly affluent and middle class lifestyle. The people who lived in Sydney, for example, came from gentrified inner suburbs or wealthy eastern ones. One programmer, when asked why he worked at home, replied from the comfort of his fully equipped kitchen, with a gesture toward his courtyard garden, 'Who would not want to work here?'

It is unusual for sociologists to be interviewing people very like themselves rather than some socially disadvantaged group. But our sample of new technology outworkers would have to be classified as professional, in stark contrast to traditional outworkers. This applies to almost all the word processors as well as the programmers if, following sociological convention, the status of married women is derived from that of their husbands. There was something very familiar about this self-description by Alex:

> When we look at magazines and papers, we see we are just another couple of yuppies—we are trying to start a family in our 30s, we have lived together for ten years and on our tenth year anniversary we decided to get married and we had a great party.

Seventeen members of the sample, including twelve male programmers, said that they had a spouse who also worked at home, four of whom were described as housewives. Among the others were six programmers, two architects and a photographer, while three of them were assisting their partner's business, doing general office work. Almost two-thirds of the sample (21) said that they were fitting their working hours to those of their spouses.

Conclusion

In some respects perhaps, the homeworkers we talked to could be said to approach Toffler's vision of the nuclear family as the new work collective. But it is important to recognise the specific characteristics of the occupational groups we studied, and the reasons why their experiences are not susceptible to generalisation. Looking at programmers, what are the conditions for their successful homeworking? Firstly, they have the considerable financial resources necessary to purchase very expensive equipment and, secondly, they have a skill which is very much in demand. Their strength in the labour market is a product of the well documented chronic shortage of programmers and systems analysts in Australia (Daniel, 1986). The labour process literature has been vitally interested in the question of struggles for control over work, but it has tended to study mechanisms of control that are internal to the workplace. For many groups, however, their ability to control their work depends critically on their position in the labour market. The skills of programmers and their continued control over both conception and execution in programming give them unusual labour market strength. Programmers may well prefer self-employment and contracting to employee status because of their ability to exploit the skill shortage in their area. The strength of their bargaining position more than compensates for the loss of industrial protection attaching to employee status.

If they lack particularly sought-after skills, workers have an alternative way of seeking to protect their pay and conditions in becoming employees in a unionised workplace. Organisation in unions is another way of increasing bargaining power in the labour market. Interestingly, our programmers were not only non-unionised but also explicitly hostile to collective organisations such as associations or unions. One programmer, for instance, saw them as 'a refuge for the weak'. This probably reflects programmers' professional status and class position, and their already privileged labour market position.

The financial conditions of entry into self-employment in programming make this a relatively small and privileged group. And their very success in developing software may in fact, at the same time, be eliminating the strong demand for their labour. Software work, a relatively young occupation, has already been fragmented—split between management, design and execution—and this is enabling the rapid substitution of machines, systems and software for human labour (Kraft and Dubnoff, 1986). Consequently, the relatively privileged position of many software workers may be a temporary condition. For these reasons it is not at all clear that we can draw any general conclusions about a trend towards homeworking from their experience. Rather, it is important to periodise the process of technological change.

Word processors have skills which are not as rare as those of programmers, yet they too were hostile to unionism. We can speculate that in part this reflects their class position, inasmuch as this is derived from that of their husbands. In addition, if we think of their work as typing done on very expensive equipment, then it is work that has traditionally been non-unionised, both in the private sector generally and at home. Word processors working at home are competing with 'temp' typing agencies, and this is the least unionised sector. Competition is not likely to decrease precisely because it is work done by women, and women are in an inferior labour market position generally, So that, whereas the programmers we spoke to had an apparently endless supply of work, the word processors complained of a highly irregular supply and therefore of irregular earnings. Some feminist critics of the labour process literature point out that, in focusing on the dynamics of the workplace itself, it has tended to ignore how sexual and ethnic divisions in the labour market have effects in the workplace (Cockburn, 1985; O'Donnell, 1984; Game and Pringle, 1983). It is important to recognise that the skill involved in typing and word processing is far from negligible, but that it is systematically undervalued because it is identified as women's work

The sociological literature on women workers has emphasised the ways in which women's position in the labour market is influenced by the fact of their primary responsibility for domestic labour within the family (Wajcman, 1983). Explanations for women's outwork have emphasised the same point. In fact, women's outwork has been shown to reinforce the sexual division of labour within the home and in the labour market. Our study shows that this is also likely to be the case for most women working with new technology at home.

By contrast, the men who worked at home as programmers came closer to Toffler's vision of a more flexible lifestyle, integrating home and work. But, as we have seen, rather than contributing to a less rigid sexual division of labour within the home, they tend to become obsessive about their work. Toffler and others who write about new forms of homeworking tend to write about it as though it is the key to work in the future. All the literature on 'post-industrial' society features working at home (Frankel, 1987). Yet such generalisations seem presumptuous when we examine concrete, historical case studies. The factors which lie behind the use of computers to work from home are different even in the two cases we studied. In this respect the existence of a particular technology, in this case the personal computer, is much less informative than an understanding of the actual work done with it, in combination with the sex of those doing that work.

We have found very little evidence to support the idea that in the case of professional couples who both work at home we will see a reintegration of home and work and a radical change in the domestic division of labour. This is not to say that many individuals will not want to work at home. There are likely to be growing numbers who would support it. But the nature of the work that people will be able to do at home, and the conditions under which they will do it, will not be determined by individual preferences. Such preferences will have to reckon with the larger forces of economic competition and restructuring with which we began this chapter, as well as the structures of sexual inequality.

5 The meat-freezing industry in New Zealand

KERR INKSON AND
PETER CAMMOCK

This chapter considers the New Zealand meat-freezing industry as a case study in labour process analysis. First of all, the replacement, from 1932 on, of the traditional system of 'solo butchering' by the 'chain system' of automated disassembly is considered as a prototypical example of the Braverman (1974) hypothesis of deskilling of work through the imposition of 'scientific management'. Secondly, historical data concerning events subsequent to the introduction of the chain system are used to support the critiques of Braverman which emphasise labour process as the outcome of collective and individual resistance and struggle by workers. Thirdly, a usually ignored variable is considered—the subjective expectations of, and response to deskilling by, the workers involved, their work orientations, attitudes and behaviour. Finally, the various strands of the paper are drawn together in consideration of the key influences affecting the labour process in the meat industry at the present time.

The study

The basis of this chapter is research conducted sporadically by the authors over the past ten years. A study of the introduction of the chain system in 1932–33 was based on historical accounts of the major dispute of that season and the issues which catalysed it, set against the labour process analysis perspective of Braverman. Newspaper articles from the period, company biographies, correspondence from union files and retrospective interviews with participants were all used to gather data. This study is described in detail elsewhere (Inkson and Cammock, 1984). A second case study relating to current mechanisation of pelting operations was based on observations in the Waitaki International company, and unstructured interviews with managerial staff, union officials and workers (Cammock and Inkson, 1985). Other relevant data came from a cross-occupational study of job satisfaction of manual workers (Inkson, 1977a), and from an attitude survey of freezing workers aimed at assessing their orientations to work (Inkson, 1977b), aspirations (Inkson, 1978), attitudes to industrial relations (Inkson, 1979a) and union involvement (Inkson, 1980), and from a participant–observer study of worker adaptation and inter-group relations in a 'chain' situation (Inkson and Simpson, 1975). An Australian perspective on labour process in the industry, focusing mainly on trade union reactions, is provided by Willis (1985).

The introduction of the chain system

To tour a modern export meat-freezing works is an enlightening, though somewhat gruesome, experience. The first things that someone who hasn't been there before notices are the sights and smells of the product: bright spouting blood, stripped shining flesh, half-severed heads dangling grotesquely from swinging carcasses, steaming entrails and internal organs tossed casually into shining steel trays. There's a quality of hygiene nowadays that wasn't always there—no grubby back-street factory this, but something nearer to a mechanised operating theatre, with lots of plastic and stainless steel, water pouring around to flush away impurities and dazzling white uniforms for the staff. But the noises are of the factory: pounding motors driving the clanking production lines which move the product relentlessly on. Amidst it all are the workers, busily pulling, cutting, trimming, washing, grading; fingers working, knives flashing, over and over again, the same precise, practised, drilled sequence: a job cycle over in seconds, active but mindless and automatic. A life's work might be to cut a fresh throat, or trim another cut of lamb, every seven-and-a-half seconds, eight hours a day, for 40 years.

It was not always so. The old, faded black and white meatworks pictures show slaughtermen in black singlets, each in his own wooden pen, each an artisan as the shearer still is, each able to slaughter and dismember an animal from start to finish. The speed would be set by 'the clock', the union delegate, his experience and his built-in metronome pacing all the others to a kill of a hundred each per day—far above the per man kill of today's 'efficient' mechanised chains—and woe betide the man who got ahead of him. There was artistry and precision in the work: many men aspired to be butchers, but few succeeded, and those who did were cocks of the works, revered for their skill and above the others in status and in pay. They controlled the union, controlled entry to their jobs and in some respects controlled management. Their control came from their skill.

Meanwhile, management in organisations such as the English meat-processing firm, Borthwick's, chafed and railed against the control of the butchers. The situation in the works was not only expensive but from management's point of view, *unnatural*, with control in the wrong hands, and the manager's supposedly divinely granted prerogative ('management's right to manage') clearly denied. On the capitalist's behalf, says Braverman, the manager strives always to impose control. In this case it was imperative for management to break the grip imposed by the monopoly of the solo butchers over their scarce skills, and their collective strength in imposing their will. In the early 1930s, a fortunate (for the employers) coincidence of external events gave them their chance.

First of all came the Depression, which brought hard times, low prices, the threat of lost sales and a growing pool of unemployed men desperate for even scab jobs. Then there was the existence overseas of the chain system, with a new promise of industry efficiency, the Ford-type mechanisation of meat processing, already extensively in use in America. When, in 1932–33, the meat workers went on strike to protect themselves from threatened pay cuts, management, who in some works had previously stored 'chain' machinery in anticipation, quickly set up the new system. The solo butchers were replaced by a 'chain'—a slow-moving mechanical conveyer from which the carcass dangled, and which pulled it past a line of workers, each of whom completed his own operation upon it until, between them, they had done all the work of the solo man. Head, hoofs, liver, entrails, heart, tongue, hide and other organs were stuffed into appropriate containers and chutes and taken to other departments for further processing. It was like a car assembly line in reverse—if you like, a 'disassembly line'.

Since only minimal skill was required for each job, unemployed scab labourers were recruited and quickly learned enough to get the kill quickly up near its maximum. Lacking both the premium skills and the collective strength of the slaughtermen, they could be paid at modest rates (in one works, about 40 per cent of slaughtermen's pay). The meat workers, believing that unskilled men could never create an adequate product, and hoping against hope that something would go wrong, picketed forlornly outside the works. But this was to no effect, for in most cases the new system was a roaring success, quickly approaching the old in terms of labour productivity and far outstripping it in terms of reduced processing costs. A few butchers were induced back into the plants to work as trainers or as ordinary labourers; others moved on. The meat workers' union was broken and, on 12 May 1933, the six-month strike was abandoned.

The story of the introduction of the chains is a prototypical *vignette* of the deskilling phenomenon. The complex craft of butchering, which had taken those men capable of learning it years to acquire, was fragmented into 30 or 40 discrete, simple cuts. The planning and pacing of the work passed from the hands of the workers into those of the managers and engineers who supervised the chain system: conception thus became totally separated from execution. Management became the major repository of knowledge and expertise in the meatworks and the workers became automatons whose ideas and skills beyond their few simple cuts were irrelevant to the logic of the production process. The price of labour (and therefore the wage which the worker took home) was reduced from that of a scarce and complex skill, monopolised by a cohesive union, to that of simple unskilled toil which could be performed by any man who happened along in need of a job. The workers were no longer proud and powerful artisans but dominated elements in a bleak mechanistic production process.

The case also illustrates rather nicely the duality of managerial motives for the deskilling of work: the motive of efficiency and the motive of control. The principles of scientific management, as developed by F. W. Taylor (1947) and his successors, are geared above all to efficiency. Specialisation, for example, reduces training time and labour cost. The human body is viewed as a machine of great productive potential which will, however, be wasted if unthinking workers are allowed to divert its energies to their own slipshod methods. Management alone has the overview and the brain power necessary to devise the total system into which the worker–machine becomes, in management's mind at least, a production element, an extension of the machine, a *thing* to be used as are other things. Yet at the same time management recognises the unfortunate tendency of the 'thing' to take on a life of its own, particularly when collectively organised. Taylor, for example, saw 'systematic soldiering' or restriction of output as a major threat to the efficiency of work. Therefore the second managerial motive is that of control: only through management control over labour does real efficiency become possible. In the pre-chain era, the union, both through its monopoly of skill and through the 'clocks', exercised control. Afterwards, management ruled by division in seizing the very heart of the production system from the men and embodying it in the new mechanical system. Our evidence suggests that at the time the wish to break the union was a major managerial motive for introducing the new system (Inkson and Cammock, 1984).

Critique: subsequent change and worker resistance

On the surface, the events of 1932–33 provide a clear and powerful illustration of the Braverman analysis, particularly the use of mechanisation as a means to subjugate labour and control labour power. The use of machines to deskill work, the decimation of complex skills and labour value, the separation of conception from execution, the managerial drive for control: all are represented in the case. However, adherence to the Bravermanian view as a paradigm for the development of labour process in the industry can be maintained only if we confine our attention to that single, one-off transformation, and ignore the historical sweep of technical development, control and labour relations in the industry, particularly subsequent events. Braverman's analysis suggests that deskilled jobs can subsequently change only by becoming further deskilled. Yet the most cursory observation of meatworks and the working conditions of meat workers at the present time (indeed at any time in the past 30 years) shows us that, although the work process is essentially the same as that introduced in the 1930s, the work is no further deskilled, and in some respects may even have become reskilled.

For example, the status of 'knife hand' (fully trained butcher, able to work effectively and quickly at any point in the chain) is admired and sought after, and requires extensive training and experience. 'Rover' knife-hands (moving from work-station to work-station as required) are needed to provide flexibility and to counter absenteeism. At many works there are informal job rotation systems. The butchers' group remains the largest and most powerful in most works. In addition, changes in consumer demand, hygiene regulations, electronic and other technological developments, and more sophisticated conceptions and techniques of management have created within the meatworks a wide range of occupations whose skill, however defined, is clearly at a relatively high level. These occupations include graders, inspectors, maintenance engineers, electricians, technicians, industrial engineers, designers, trainers, and specialists in stock control, finance, industrial relations and marketing. These people make the works a more complex organisation than it was 50 years ago, and represent the outcome of a range of forces in industrial development—for example, technical change, consumer sophistication and management science—which exert pressure on the labour process in a manner directly opposite to the pressures described by Braverman.

The point is also made in the introduction to this book (Chapter 1) that the position of labour arises from a combination of external 'objective' forces destroying individuality, and worker responses and resistances to these pressures. While management secured an important victory over labour with the introduction of the chain system, collective struggle and resistance by meat workers and their unions in the post-war era had, by the 1970s, led to a far more even distribution of control between management and workers than a straight application of labour process principles would lead one to expect. If chain technology is against them, meat workers have many environmental factors on their side in the struggle: the interdependence of the workers in the new system, the essential perishability of the product, the fact that animals lose their value quickly if not killed while in prime condition, the lack of competition in the industry and the concentration of work in a relatively small number of large export killing works which could easily be unionised. Additionally, as we shall see, when collective struggle meets with success through strong unionism, then individual and small-group resistance also becomes more feasible.

The meat industry is one of many in which the 'master plans' of management, based on rationalised formal control, have, in practice, gone seriously awry. The meat unions, encouraged by external forces such as the Labor governments of 1936–51 and the assured markets and effortless prosperity of the 1950s and 1960s, have rebuilt a formidable position of strength within the industry. Further, as in many rationalised industries, the act of *forcing* workers to do routine monotonous jobs has led to many individual resistances, for example, absenteeism, tardiness, indiscipline, horseplay, safety violations, theft, sabotage, group restriction of output and passive resistance to authority. The combination of collective and individual resistances led eventually, despite the apparent domination of chain technology, to a situation in which many *managements* felt demoralised, dominated and lacking in control (Inkson, 1979b).

The practices of successful modern management in the industry are therefore likely not to be those of coercion (for the workers tend to have too much power to be coerced) but those of 'responsible autonomy' for workers (Friedman, 1977b), and those of negotiated change. In many cases, meat workers and managers have unwritten agreements as to the real exercise of control in the workplace, for example, regarding the quality standards to be expected, the extent to which workers may be harassed by the supervisor, or even whether, provided performance targets are met, workers need to be supervised at all.

The subjective nature of deskilling

A point of considerable interest, but one typically neglected in labour process analysis, is the *subjective* nature of the process—the experience and judgement of the worker. How does the worker *feel* about the imposition of technological and management change, how does change relate to his or her work aspirations, and how are attitudes shaped by the experience of work?

Like much Marxist theory, labour process analysis tends to treat worker attitudes as somewhat irrelevant. It is argued that features such as the existence of a class society, the imposition of rationalised management control and the degradation of work are 'objective' facts, determinable by the outside observer. The fact that the worker may have little consciousness of class, and may not feel that his or her work is especially degraded, is of no consequence. It is, after all, easy enough for the capitalist socialising institutions to induce a 'false consciousness' by workers of their true conditions.

On the other hand, managerially oriented sociologists (those concerned with what the introduction to this book calls 'sociology in work') have a long tradition of interest in workers' own perceptions of their work, their attitudes and job satisfaction and, by implication, their mental health and completeness as human beings. Apart from the 'objective' degradation of work which may be entailed in the labour process, workers may 'subjectively' feel themselves to be 'victims', in a direct, personal way, of management or of technology.

Studies of assembly line technology have drawn attention to the low job satisfaction, poor self-esteem, high aggression and poor mental health which workers frequently display in this situation (e.g. Blauner, 1964). The 'alienation hypothesis' which explains such responses is based on a psychological view of human beings as having 'higher order' needs such as achievement, autonomy, self-esteem, variety and self-fulfilment, which are obstructed by the reality of job conditions. The alienated worker not only feels dissatisfied, disconnected and fragmented, but reacts 'pathologically', with aggressive anti-managerial behaviour.

Underlying both the Braverman labour process degradation hypothesis and modern concerns about worker alienation from work is an assumption that there is a pure ideal form of manual work, of craftsmanship, in which the process of construction is integrated in a completed form, and all skill and expertise is vested in the worker. For example:

> Craftsmanship as a fully idealised model of work gratification involves six major features. There is no ulterior motive other than the product being made and the process of creation. The details of daily work are meaningful because they are not detached in the worker's mind from the product of the work. The worker is free to control his own working action. The craftsman is thus able to learn from his work; and to use and develop his skills and capacities in its prosecution. There is no split of work and play, or work and culture. The craftsman's way of livelihood determines and infuses his whole being. (C. Wright Mills, 1951: 220)

Comparison of modern occupations with the ideal reveals that the skilled trade engineer, for example, comes closer to it than does a process worker. But even the engineer must normally work to instructions from management and integrate his or her efforts within an overall production framework. At a higher level of craftsmanship are 'home' craftspeople such as potters and weavers, who control not only the manufacture of the entire product (not just a component) but also its design and sale. A study of potters revealed that, in many respects such as integration of life and work, high job satisfaction, self-development and esteem through one's work, such craftspeople come close to the ideal (Inkson, 1987). But even their work is degraded in objective terms by the crude necessity of economic survival in a competitive society and the consequent necessity of arduous toil to produce the standardised functional products which consumers seek. Here it is not the manager but the price-conscious consumer who exercises 'control' over the worker. Subjectively, too, economic insecurity mitigates the pleasure and fulfilment of the craft. Even in 'ideal' conditions, therefore, the ideal of craftsmanship is tarnished by the reality of economic necessity.

What of meat workers? To what extent was the solo butcher of the past a true craftsman? He (for he was always male) certainly commanded a complex skill, completed a meaningful cycle of production and gained identity and status in the community through the nature of his work. On the other hand, the duplication of the same cycle of activities fully 100 times in a day is as repetitive as the work in many modern process jobs. The work was also physically arduous: the butcher processed the animal on the floor, holding it between his legs, in an awkward position which after a few years tended to result in his developing physical deformities. Most ex-solo butchers we spoke to remember the transition to the chain system as being to less arduous rather than to less interesting work. The ideal of 'pure' craftsmanship seems unlikely ever to have been characteristic of meat work. Just as Braverman's view of the enslaved modern production worker may have been rather too pessimistic, so too was the view propagated by his analysis of the pre-industrialised craftsman, in some cases, excessively romanticised.

The work of modern meat workers is certainly routinised. Freezing workers, according to the evidence of the attitude study, have low job satisfaction (Inkson, 1977a), and see 'monotony' as the major negative feature of their work (Inkson, 1977b). They show high levels of absenteeism, aggression, indiscipline, sabotage, accidents and strike action. Both they, their managers and their representatives see the nature of their work as a major predisposing cause of conflict (Turkington, 1976). On the face of it, a link between deskilling, dissatisfaction, alienation and aggressive job behaviour appears at least plausible.

However, other evidence suggests that, although meat workers have negative attitudes to their jobs, these are integrated into a framework of expectations and adaptations, often enabling the worker to cope adequately with such feelings. Meat worker are generally 'instrumental' (Goldthorpe et al., 1968) in orientation. That is, they see work as labour, as something of value not for itself but only for the financial benefits it provides. In the attitude survey, in answer to the question, 'what is it that keeps you here?' nearly 90 per cent of respondents mentioned wages. For over 60 per cent this was their only response (Inkson, 1977b). This valuing of wages was linked to the desire for an affluent lifestyle, for possessions, holidays, leisure activities and a secure family life (Inkson, 1978). The meat worker builds his dignity through his off-the-job activities, and *himself* degrades his work to a subsidiary status in his life.

Meat workers, on the evidence of the above survey, do not see either their work or themselves as being 'degraded' or 'subjugated'. They have their own, eminently sensible, ways (as have any workers in their predicament) of coping with it, and of making the best of it. Absenteeism in the form of 'sickies', for example, may be a good way of breaking the pressure. Meaning, interest and variety denied by the technology may be substituted by meaning, interest and variety gained from social interaction and from intra- and inter-group activity. A 1975 case study documented rich patterns of interaction, ritual and inter-group rivalry in a meatworks, enabling individuals to find worthwhile social meaning in their working lives (Inkson and Simpson, 1975). The same study showed how workers can adapt psychologically to objectively degraded work by putting their own definitions on its meaning. A group of chain labourers, for example, adapted to their arid jobs by focusing on the relatively high intrinsic rewards, while product inspectors, less well-paid, emphasised their status as trained men who had the prerogative to interrupt production.

Over the years, meat workers have bargained for good economic rewards for the work they do and have been concerned chiefly to consolidate these gains. Objectively the work may be degraded; subjectively the worker regards the degradation as a matter of little importance except for any effects it may have on his wage packet. Though control over the organisation may be enhanced by 'worker participation' and control over the job by 'job enrichment', the meat worker spurns both as distractions from his major goals (Inkson, 1977b, 1979a). Similarly, union activity, though an important source of resistance to managerial control, should in the worker's eyes be directed mainly towards maximisation of economic rewards (Inkson, 1979a). The union, of course, may have a more political agenda, and may pursue the struggle for control with greater vigour on the worker's behalf than he would on his own.

Although, at some global level of analysis, theories of monopoly capitalism, deskilling, degradation and indeed collective resistance, may be valid, nevertheless the impact of management and technology on the individual person and his or her individual response may be quite different. Labour process analysis, after all, is analysis not of workers and their lives, but of the 'labour process'. Like much sociology it can therefore be guilty of its own degradation of labour, or rather of labourers, viewing the individual as an inter-mixed constituent in a homogeneous labour mass, rather than as a person. Our own perspective, derived from psychology as much as from sociology, and from 'sociology in work' as much as from 'sociology of work', has its own limitations: there is a bias to management, an inbuilt assumption of some degree of harmony between managerial and worker interests, an unwillingness to look outside the factory to the broader societal and economic issues shaping and constraining the relationships and conditions of the workplace. On the other hand, this approach pays closer attention to the micro-level of organisational analysis, accords proper legitimacy to the subjective viewpoint of the involved members of the workforce and suggests perspectives for improving working life for working people (for example 'job enrichment') without the necessity of fundamental redemptive political change.

Technological innovation in the 1980s

The introduction of the chain system, while cataclysmic in impact, did not finish forever the development of technology and labour process in the meat industry. The chains run only one part (albeit the central part) of a meatworks. Over the years, further incremental automation has taken place in many areas of operation. Some of these changes result from the efficiency and control motives described earlier; in other cases they are managerial responses to external pressure. In modern times international market conditions for meat have changed dramatically, and the worker–machine systems of the industry have had to respond. For example, slaughterboards have been reorientated so that Moslem slaughtermen face Mecca during the killing of lambs for the Iranian market. Processes and jobs have been added to the chain, and equipment redesigned, to meet modern international hygiene requirements. Increasingly, in a world with a growing meat surplus, control is exercised not so much by management for the sake of capital accumulation, but by overseas consumers for the sake of price, quality and other product features which they value.

In New Zealand, a new rationalisation is taking place in the meat industry. It is clear that the industry has an overcapacity in relation to market demand and that in the next few years as many as 20 per cent of the country's export meatworks will have to close. Some have already done so, the latest, at Whakatu near Hastings, at a cost of some 1500 direct jobs. The works which are most under threat are those which have the most out-of-date plant or are the least efficient and profitable. Within each works, a new communality of interest is created between management and workers as each gains by enabling the works to survive. At the most basic level this may mean no more than decreased worker resistance when managers attempt to enforce increased productivity or quality, or reductions in wages. In other cases, managers and workers may negotiate actively to accommodate the others' interests in a package for mutual benefit or mutual survival.

As a case study in the rationalisation of technology in the modern meat industry, we examined the introduction of a new process, the ring pelting machine, in one meat company during the early 1980s (Cammock and Inkson, 1985). The pelting machine represents not just rationalisation but automation. Whereas the chain style mechanises only the transport of carcassess (all the butchering is still completed by hand), the pelting machine automates a set of manual process operations, by stripping the pelt from the animal mechanically. Up to fifteen jobs per chain are replaced by two jobs operating the new machine. Labour is not deskilled or degraded; it disappears altogether (apart from the machine operators). While there are quality problems with the machine, the potential cost savings to management are considerable. Management will, of course, argue that in the modern oversupplied competitive environment companies which fail to capitalise on technological developments will simply go out of business, a point of view against which it is hard for the union to argue.

The process and outcomes at the company we studied were instructive, and provide a good contrast with the events of 1932–33. After protracted discussions it was agreed to install pelting machines in two works. Union officials negotiated the introduction defensively, in the words of one of them, 'resisting and relenting', because they could see the competitive pressures on the company, the inevitability of technological change and the possibility of large-scale redundancies in the longer term if they resisted without relenting. Interestingly, in terms of control over the labour process, automatic pelting clearly creates redistribution of control, not just from workers to management, but also from workers to other workers. Maintenance staff responsible for the operation of the new technology developed a high level of expertise in its mechanical, hydraulic and electronic components. This expertise was shared neither by management nor by shop-floor employees. The maintenance men became the key controllers of a resource on which the entire work was dependent.

The slaughtermen displaced were not only to be relocated in alternative jobs elsewhere in the works, but were to have their previous rates of pay as butchers maintained. The brutal dismissals of 1932 have their counterparts among the redundancies at Whakatu (though at least modern workers receive redundancy pay), but not among the reorganisations of work at facilities located at Stoke and Picton, where the pelting machines have been installed. Nor is there any suggestion of class struggle in the responses of workers or union to the machines. If there are victims of the change, they are the young unemployed people who might otherwise have found employment in the works. But essentially workers, union and management have closed ranks to benefit those who already have their security invested in the works. As for the language of labour process analysis—'deskilling', 'degradation', 'control', 'separation of conception from execution'—it seems irrelevant to the issues negotiated on.

Conclusion

Our conclusion is that at the micro-level of a single major change to a new 'scientific management' technology, at a point in time over 50 years ago, the Braverman analysis makes a good fit. However, at the macro-level of the whole industry over a long period, the proposed unilinear sweep toward totally deskilled work and absolute managerial control is not in evidence. Rather, the introduction of chains is just a rather dramatic part of a shifting pattern of power-play, resistance and consolidation by both sides of the industry. Organisation of the labour process is too much influenced by extraneous factors other than the capitalist drive for capital accumulation—for example, external economic, and market pressures, opportunities for technological development and changes of the public view of acceptable business behaviour, all influence managerial goals and responses. Workers, too, as we have seen, have their own powers and goals which lead both to resistance and to adaptation, and thereby transform the labour process in directions which might not be anticipated by objective analysis.

6 Quality circles in the manufacturing industry

RICHARD DUNFORD AND
PETER McGRAW

Labour process theory gave voice to a disparate range of criticisms of the limitations of most established variants of organisation theory. For example, the human resources advocates of job enlargement, autonomous work groups and the humanisation of work were seen as sociologically naive, their analysis lacking any comprehension of phenomena of class and power. For a decade, post-Braverman labour process theory gave a new coherence and direction to sociological analysis of workplace relations. However, labour process theory was painting itself into a corner. An approach which had initially provided a critical reassessment of efficiency-based explanations of organisational phenomena threatened to become a new orthodoxy of an equally vulgar sort. The emphasis on control as *the* explanation for management actions restricted labour process theory to a rather simple determinism typified by the comment that '. . . the most efficient technology (in terms of the maximum production) will be chosen only if it is compatible with securing the maximum control over worker behaviour' (Clegg and Dunkerley, 1980: 343). The phenomenon of management was treated as being of little interest because it was 'simply regarded as a structural mechanism through which the changing pattern of organisational requirements necessary for continual capital appreciation is realised' (Reed, 1984: 276).

However, in recent years there has been a change—a loosening of the orthodoxy—which has reinvigorated labour process theory. Central to this new direction has been the notion that neither the grounds for reorganisation, nor the actions of management, nor the outcomes of such actions are known *a priori* (Knights et al., 1985; Knights and Willmot, 1986). Implicit in this is the importance of empirical case material which seeks to establish a greater understanding as to just what reorganisation of the labour process involves in specific situations.

The focus of this chapter is the analysis of the use of quality circles, which are typically described as groups composed of individual workers who meet regularly in paid time to identify and analyse problems in their immediate workplace with regard to factors such as quality, productivity and safety. The circle is then supposed to assess alternative courses of action, make recommendations to management and, where possible, oversee the implementation of changes. Each circle has a leader, usually the supervisor of the section from which the members come. They are supposed to make decisions collectively with the leader facilitating discusssion rather than issuing orders.

The dominant discourse surrounding quality circles is the human resources approach of most quality circle advocates (e.g. Dewar, 1980a; Thompson, 1982; Robson, 1984; Hutchins, 1985). The mutual optimisation of efficiency and participative workplace practices is assumed. The overall objectives of quality circles, as specified by the author of a pre-eminent handbook, are to reduce errors and enhance quality, inspire more effective teamwork, promote job involvement, increase employee motivation, create problem-solving capability, build an attitude of problem prevention, improve communication in the organisation, develop harmonious management–worker relationships, promote personal and leadership development, develop a greater safety awareness and promote cost reduction (Dewar, 1980b: 2–4). Some writers also tend to be highly evangelistic. For example, Hutchins asserts that, 'Quality circles are the most exciting and profound approach to have been established in the world since the advent of scientific management' (1985: 2). Nor are these writers reticent about attributing a wide range of outcomes (presumably desirable) to the operation of quality circles:

> quality circles produce dramatic results for an organisation. They improve morale, increase a sense of loyalty to the organisation, and foster a sense of teamwork among employees who participate. They improve overall productivity of the organisation, and they improve the quality of the product or service. And they reduce grievances, lost time, accidents, scrap, attrition, absenteeism, and tardiness. In short, they solve problems and save the organisation money. (Thompson, 1982: 11)

This interpretation of the significance of quality circles can be contrasted with a labour process perspective provided by Grzyb (1984), for whom a quality circle program is one of a number of forms of recollectivisation of the workforce which '. . . represent attempts by management not only to boost productivity, but also to erect new structures of control over the workforce and the labour process without by any means abandoning older methods for achieving such domination' (1984: 144). Manwaring and Wood speak a similar language but present a modified labour process perspective, arguing instead that such practices 'are not necessarily cosmetic or necessarily manipulative', but rather that they are 'minimal joint creations within the context of capital's domination' (1985: 187).

Just what quality circles are, is itself a contested matter. Even the way they are to be treated within labour process discourse is not unambiguous. In the following section the use of quality circles in two organisations is examined.

The study

The study took place in Reckitt and Colman Foods, and Reckitt and Colman Pharmaceutical Division, both divisions of Reckitt and Colman Australia Limited, which in turn is a wholly owned subsidiary of the British company Reckitt and Colman PLC. Data was collected from a combination of semi-structured and structured interviews plus archival material (company files, quality circle records). In total, 44 interviews were held, 22 in Foods, 20 in the Pharmaceutical Division and two with union organisers. The two managing directors were interviewed as well as four managers, ten supervisors and 26 worker members of the circles. The gender breakdown was 31 male, thirteen female, all of the latter being in the worker member category. The study is reported in full in McGraw et al. (1987).

Quality circles in Reckitt and Colman Pharmaceutical Division

Reckitt and Colman Pharmaceutical Division manufactures a range of products including analgesics, antiseptic and disinfectant and also packages imported ethical (prescription) drugs. The factory is divided into a number of discrete automated or semi-automated production lines. The work of operatives is essentially concerned with loading ingredients into mixers which feed the lines, monitoring and servicing the lines and packaging the finished products. The division employs 45 non-award staff (management, senior clerical staff, secretaries, supervisors) and 150 award staff. The majority of the latter are members of the Australian Workers' Union (120), with the Amalgamated Metalworkers' Union (13), the Association of

Professional Scientists of Australia (7) and the Miscellaneous Workers' Union (5) accounting for most of the remainder. Wages and conditions for employees are determined by award and supplemented by a site agreement which is negotiated jointly by the unions and management on an annual basis.

In October 1980, as part of a major reorganisation within the Reckitt and Colman group, a new managing director was brought out from the United Kingdom to head the division. Around the same time the group training manager was having discussions with a firm of management consultants who were at that time promoting quality circles. A presentation was arranged at which the consultants outlined the benefits said to flow from the use of circles. In January 1981 a decision was made to introduce them into the division and, following a period of consultation with site unions (which raised no fundamental objections), the program commenced in June 1981.

The introduction of the program seems to have owed a lot to the current concerns of the new managing director who, shortly after his arrival, began to express dissatisfaction with the current state of worker–management relations on the site. The vision presented by the consultants hit a responsive chord. Quality circles struck the managing director as an effective catalyst for certain changes that he deemed necessary. Specifically he expressed a wish to improve communication between senior management and shop-floor workers, to establish some regular basis for dialogue with the shop-floor which did not involve trade unions, to develop the grievance-handling skills of the supervisors (who were currently referred to as 'cardboard cutouts', such was their lack of action in such matters) and to develop a means of encouraging and legitimating the 'creative engagement' of shop-floor workers in the process of identifying and solving problems connected with the production process.

The response of the supervisors to the quality circle program was initially one of uncertainty and resentment, fuelled by their compulsory involvement as circle leaders. However, this subsided as they saw the operation of circles as part of a more widespread change in management 'style'. One supervisor commented, '(previously) we had no contact with senior management . . . The [managing director] has changed the concept of management here; he keeps people informed and comes down to the floor and talks to people. People know who he is and what he's trying to do.'

Similarly, the workers, with few exceptions, were enthusiastic about the idea, in particular responding well to the idea that ideas from the shop-floor be encouraged. Whilst one or two complained that quality circles were yet another management device for getting more out of the workforce, the vast majority saw the changes as 'sensible management'. Within the framework used by these workers, management overtures designed to utilise aspects of their tacit knowledge in this fashion were seen primarily as timely recognition of insights that previous experience had taught them to expect to have ignored. It was a case of 'about bloody time' rather than resistance to the scheme.

Within the first two years of the program, eight circles became established, involving 60 per cent of the workforce as members. A number of changes occurred within the division associated with the existence of circles. There were concrete changes to workplace practices and procedures which developed from quality circle proposals. On the 'Steradent' production line a new washbay was designed and built, dust extractors installed, the line redesigned for easier access and a noise reduction program initiated. In the Dry Mixing area the dispensary was modernised, loading platforms installed and racks altered for ease of access. Changes such as these occurred throughout the factory. Taken one at a time they rarely appeared to be changes of great moment; however, the cumulative effect of these changes was to reinforce in the workers' minds that their ideas were considered and acted upon in ways which concretely affected their immediate work environment. Both management and workers attested to the improved communications through the organisation whilst management was heartened by the significant improvements in productivity, profitability and industrial relations. The managing director expressed great satisfaction with the effect on supervisors ('unrecognisable compared to the supervisors we'd started with') whilst the organiser for the major union on site viewed the developments favourably, particularly due to the fact that he was much less often involved in the resolution of grievances which he saw as minor and irritating, thus freeing him for 'more important' items.

However, by mid-1984 only one circle remained active. There seem to have been a number of factors acting in concert to bring about the demise of circles. Firstly, the circles ran out of ideas for new projects. Secondly, approved projects were not being completed because implementation was usually dependent on the engineers who in turn had to fit this work in along with normal maintenance tasks. Thirdly, the pressure of production made it difficult for supervisors and workers to fit in quality circle meetings. Fourthly, the supervisors resigned *en masse* as circle leaders. They were feeling the effect of production pressures and were uncomfortable in a role which required them to act as progress chasers. Finally, the manager primarily responsible for liaising with circles, the 'facilitator', was not provided with sufficient resources to deal with the demands of all eight circles simultaneously.

The most striking aspect of the responses to the demise was the different views of top management and workers. The demise was of no particular concern to top management, as in their view the circle program had achieved what it had been set up to do. The managing director's view was, 'I think we were beginning to operate on the law of diminishing returns. We were flogging a dead horse with quality circles, they had actually achieved, in their eyes, all they were going to get out of it at that time.' For the workers, on the other hand, the dominant view was that, through management inaction, the circles had become ineffective to the point that they ceased to function. Considerable hostility and disenchantment resulted. One commented, 'I felt really disappointed in management. I felt why should I help them with their problems? That's it, I'm out.' Another added, '[Now] I reckon they should just leave it to the top guys to work it out. I reckon I'm just here to work on the Dettol line and get my money and that is it.'

Central to an understanding of this difference is a recognition of the fact that senior management on the one hand, and quality circle members on the other, invested the circles with different meaning. For management, stimulating behavioural change was the most important aspect of the program. Quality circles were merely a convenient catalyst. Consequently, little was done to maintain them once senior management felt that the program had achieved its objectives. For the circle members, the most important aspects of the program were the physical changes that could be made to their work environment and the opportunity to be involved in decisions in this regard. For them, the cessation of the circles signalled failure, not success. Unlike management, they attached great importance to the form and not just to the function of quality circles. Their disenchantment arose from a feeling that the cause of the demise were largely matters within managerial control.

Quality circles in Reckitt and Colman Foods

Reckitt and Colman Foods produces a range of foodstuffs primarily of the convenience variety including packet meals, cakes, sauces and drinks. The plant is organised into a number of discrete areas with production taking place on semi-automated production lines. By comparison with the Pharmaceutical Division, it is labour intensive. Most of the workers are involved in the packaging of the products with a small number involved in the preparation of ingredients or tasks such as cherry sorting and hand decoration of cakes. The division employs 40 non-award staff and approximately 350 award staff. The majority of the latter are members of the Food Preservers' Union (300) with the Amalgamated Metalworkers' Union (25) and the Federated Clerks' Union of Australia (16) accounting for most of the remainder. As in the Pharmaceutical Division, wages and conditions are determined by award, supplemented by an annual site agreement.

In September 1982, as part of an effort to halt an extended period (ten years) of non-profitable performance, a new managing director was brought out from the United Kingdom. Changes took place on a number of fronts. In the marketing area, a new method of co-ordinating brand management was introduced. Under the previous system, the strategy in this regard was determined by the sum of the various approaches taken by product managers with regard to their individual products. Co-ordination was minimal. The new managing director soon introduced a system whereby all strategy matters were considered by the 'full management team' so that they could assess whether the strategy proposed for a particular brand was compatible with the broader strategic directions of the division. In the financial area, a new method of costing was introduced in an effort to more accurately identify the cost components of each major manufacturing process and in turn the profitability of each segment of the division's operations. The result of this was a culling of 'non-contributing' areas. Packing of some products was subcontracted, one product range sold to a competitor and another simply deleted. Purchasing policy also radically changed. Previous policy was to always buy raw materials at the lowest price despite associated quality problems, poor yields in the factory and declining sales. The rescinding of the policy led to purchase of higher quality materials, higher yields which more than compensated for the higher prices paid and a halt to the sales decline.

The new managing director inherited a site where management–worker relations were often poor and dominated by a centralised, authoritarian style of management. This he saw as another impediment to improving the performance of the division. His background had forged a particular management philosophy:

> I was brought up on the outskirts of the East End of London and was trained as an apprentice by the Ford Motor Company ... I worked on the factory floor and saw workers sabotage the machinery. I witnessed communist agitators including people who weren't Ford employees who infiltrated the site and would be summonsed by a shop steward when there was an appropriate opportunity to stir ... I was with the superintendent of a building when he gave instructions for someone to go out and provoke a strike because the order books were low ... It stuck in my mind that there had to be a better way of spending your working life

> than in that conflict-orientated environment . . . About twenty of my family over the years had worked in Fords and they had had a miserable work experience . . . I resolved from that time on that I was going to get into management and that when I did I was going to try to create an environment where the people doing the physical work and the people responsible for the decision-making worked together in a reasonable degree of co-operation.

In search of a less centralised and more participative style, the ranks of management and supervisors were cut by 25 per cent, bringing about shorter lines of communication and broader spans of control. Those cut were those deemed by the managing director to be 'least compatible with the management style we were going to follow'. There remained a concern at senior management level that lip service to principles could easily result where concrete practices did not exist to encapsulate, and through their operation, enforce and reinforce those principles. It is in this context that quality circles came to assume significance. However, the quality circles concept was in no sense seen as the complete answer.

The apparent early success of quality circles in the Pharmaceutical Division was known in Foods and some initial steps had been taken to introduce them there. They were thus a convenient vehicle. According to the managing director, 'We had a need to make the transition from one management style to another. If we hadn't had quality circles we would have found another way.' Nonetheless, the quality circle approach did particularly suit his belief that workers were a significant untapped source of valuable ideas and that management and worker interests were essentially in common. He stated, 'I believed if we could harness in common the energies of everyone in the division—pulling in the same direction—that we would make the company profitable.'

The quality circles relied purely on volunteers apart from the supervisors who were put in circle leader positions. The latter was a deliberate strategy by senior management as it was felt that, were this not done, quality circles could have been seen by the supervisor as a threat and this would have undermined the program. Supervisors were given a commitment by management that their power base was not going to be undermined by circles. Most of the workers responded positively to the quality circle idea, attracted like the Pharmaceutical Division workers by the implied recognition of worker knowledge. The few sceptics were as likely to suspect the motives of workers as they were those of management. One commented, 'They never used to call it quality circles. They used to call it quality circus. They thought we'd all come over here for a bludge.' One hundred workers volunteered to be circle members and eight circles were established.

Within two years of the establishment of the programs, a number of concrete changes took place which were seen by workers to be the result of quality circle activities. There was a widely held view amongst workers that before the introduction of quality circles matters were raised with management but little action was ever taken. One commented, '[Previously] it was just like talking to a brick wall and you couldn't get it done. With quality circles it can be discussed and thrashed out. They get things done.' Amongst the worker members there was an almost unanimous view that quality circles had led to improved relations with management. One worker argued, 'You're working more together. They're seeing what's going on and the problems you're having. They're coming to where you're working, so they're visible. They're not just a blank face, a nameless face.'

Management assessed the program positively, pointing in particular to the effect on the supervisors, whose transformation was seen as the critical determinant of the success or otherwise of the re-orientation of management style. Particularly pleasing to the managing director was the way in which supervisors assumed a more active role in problem solving and grievance resolution. He wished to encourage this activity, so that when a grievance occurred those involved did 'not run to Personnel, not run to the union' as an automatic reaction. To support this development, changes were made in the Personnel function. The existing personnel manager ('a great fire fighter and if there wasn't a fire he'd make sure there was one') was replaced as part of a move to give the personnel function a lower profile. This assisted an evolution to the situation where line management now deals with the unions and almost all problems are resolved at supervisor level. The personnel manager position has now been disbanded and all personnel functions are now carried out by one personnel officer.

Despite these positive assessments, by mid-1985 only two of the eight circles were meeting regularly. The metamorphosis was, however, quite different from that which occurred in the Pharmaceutical Division. In Foods there had developed a strong feeling amongst both supervisors and quality circle members that a rather rigid conformity to specified procedures actively inhibited success and that a flexible approach was necessary. According to one supervisor:

> The quicker you get to that stage the better. If you set up your own working arrangements it suits the leader better. Initially you've got to do it by the book but the quicker you move away from that regimented type thing the better the group gets on . . .

> If you run it by the book, every meeting you would find a big drop-off in people. If you make it too regimented you'll lose them.

One area in which 'following the rules' seemed to generate its own problems was in respect of the formal presentations to management. They became increasingly elaborate; 'almost a vaudeville effort' according to the production manager. One circle spent five meetings simply planning its presentation strategy. Management became concerned with this aspect of the practice. They were already worried by the fact that these highly staged presentations could be a source of divisiveness within the broader workforce. Whilst they could be a source of considerable satisfaction for both circle members and facilitator they could also serve to further distance circle members from non-members. The managing director, the production manager and the personnel officer all expressed concern that the circle members were being seen by the other employees as an elite with special access to management from which they were excluded. This added to their reasons for being supportive of deformalisation developments. They were, however, uncomfortable when some of the groups began not to meet at all. The production manager explained, 'I was becoming paranoid about the fact that meetings were not happening and started to push them . . . it seemed to me that as soon as I stopped pushing the whole thing would die.'

However, following a discussion between managers, a decision was made not to intervene. What did happen was that, in the areas where the quality circles operated, discussion occurred amongst the workers despite the lack of formal meetings. Non-circle members also appear to have become more prepared to approach supervisors and make suggestions or inquire about workplace matters. The last two improvements on the Cake line came from a group of people with not one circle member amongst them. This development reinforced for management its decision to let the quality circles evolve away from strict compliance to rules. The production manager commented, 'We realised that what we had done was just the right thing; leaving it alone. By pushing it we'd probably have killed it.' Management thus sought to legitimate the informal practices rather than to resurrect the formal practices seen as so necessary by most circle advocates.

The decline in formalisation and designated quality circle meetings is consistent with the managing director's intention that the individual workers come to disassociate success from application of a special technique. As he said:

> With hindsight I wouldn't call it quality circles. I would try not to have it labelled . . . What is important is what *I* achieved or what *we* achieved. I think there is a danger with quality circles that you get preoccupied with the hype and you actually take the credit away from the individuals . . . if you give the process the credit you are missing the whole point.

The disentanglement of technique from outcome is, however, not a simple process. Whilst in the mind of the managing director it can be conceptually made, for many of the quality circle members such a sanguine assessment of the role of quality circles has not been made. The lived experience for the worker is that the change was that quality circles were introduced and therefore it is not surprising that they are seen as a major cause of the successes. The existence of an 'objective technique' also allowed individuals to participate who otherwise might not have done so, as it presented a device for communication between various levels of the organisation that blurred the hierarchical differences which might otherwise have been a psychological barrier to participation. Workers not used to contributing ideas, not encouraged to contribute ideas, with the experience of years of non-participative management usually do become deskilled in this regard. Techniques like quality circles, with their formal procedures, provide a necessary protection, particularly in the early stages. They place the interaction on a constructively impersonal level while the ice is broken.

The decrease of reliance on formal meeting procedures is a weaning process. In the Foods Division it has already occurred to a large extent, although the process is by no means complete. Despite the apparent development of successful participation in the absence of regular meetings, all of the circle members believed that the quality circles would meet again as soon as work pressures allowed. Whilst they may increasingly seem to management to be unnecessary they still have a strong symbolic function. Any formal removal of the quality circles by management would be liable to be interpreted by the workers as evidence that they had simply been subject to the latest in a long line of management fads.

Conclusion

To characterise the use of quality circles by management as a control device would not be so much incorrect as unsubtle. Certainly there is a sense in which the ideology underlying the beliefs and practices associated with quality circles may function as a vehicle for the enhancement of a consensus doctrine. To the extent this exists, it may reduce the reliance on simple, technical and bureaucratic forms of control (Edwards, 1979). Control though the internalisation of a 'world view' has typically been associated with professional control (Salaman, 1979). However, it is quite possible that this form of control can be relevant to the analysis of the control of non-professional staff. The notion of culture control (Ray, 1986) is useful, as it acknowledges that professionals are not the only workers who may be subject to this means of control. Whilst the appropriateness of the term 'culture' in this context and its effectiveness as a means of control have both been questioned (e.g. see Turner, 1984), it does at least signify a means for attempting to influence the actions and beliefs of workers that is not directly encapsulated in other treatments of the control issue. To the extent that the functioning of quality circles contributes to the enhancement of a perception of the mutuality of interests of capital and labour, they can be seen as a control device. There is, however, much that we can point to in the study just presented which should alert us to the need not to interpret this in a simplistic fashion.

Firstly, the omnipotence of management can be exaggerated. The managers in the Pharmaceutical Division clearly miscalculated the reaction of their workers to the dismantling of the quality circle program. The ill-feeling generated by this action can hardly have been either foreseen or desired. A recent serious strike in this division attests to management's exaggerated interpretation of the effects of the program. Once such a program is established, certain expectations are also established in the minds of the workers, constraining the ability of management to manipulate the program entirely as it may wish. In a 'boomerang effect', management's commitments can come back to haunt it. Also, the program in Foods was certainly negotiated in such a way that the developments in the program were often initiated by the workers with the management left to determine how to respond. As noted previously, management intentions are not necessarily realised (Kelly, 1985; Child, 1985; Storey, 1985a). Nor should the foresight of management be exaggerated. Whilst quality circles were introduced because they fitted into a broadly based strategy of management, many of the decisions about how the circles should operate, such as the deformalisation in Foods, are revealed to have been *ad hoc* rather than part of some well thought out tactical plan. The history of the circle programs in this case suggests that the coherence of management action is often exaggerated and that, as Child (1985) and Rose and Jones (1985) suggest, much is incremental and piecemeal.

Secondly, it is clear from the Foods case in particular that the effort to reorientate management–worker relations was only one component of the reorganisation of activities. Significant changes took place in marketing, purchasing and costing practices. The rather cavalier abandonment of the program in the Pharmaceutical Division may indicate that it was not a priority concern. Management action takes place on a number of fronts, involving more than just reorganisation of the labour process. The marginal returns from action in this arena may pale next to those from changes in financial or marketing policies (Littler and Salaman, 1982; Child, 1985; Storey, 1985a). Indeed, as Child (1985) has noted, changes in the labour process may be the unanticipated by-products of a change in policy in one of these latter areas. It is important to avoid the 'just-so' fallacy: whilst certain management action may have a particular outcome, such as increased control, and whilst those actions would seem to be actions one would take if seeking such an outcome, it is important not to confuse function with cause (Dunford, 1983).

Thirdly, there are dangers in treating management as a unified entity, especially if this is attached to the notion that they are '. . . essentially unproblematic agents of capital who dispatch their "global functions of capital" in a rationalistic manner' (Storey, 1985a: 195). If the actions of a manager are interpreted as being in support of capital this causes logical problems when another manager reverses or seeks to reverse this action. Especially when these differences of opinion or action occur contemporaneously, one must conclude that diversity of opinion exists within management and that this is central to an understanding of management actions. As has been noted previously, this is not simply a matter of different views about means to a common objective, but rather indicative of differences of material interest between sections of management (Armstrong, 1986; Buchanan, 1986; Storey, 1985a, 1985b, 1986). Disaggregation of management in this sense should also alert us to give some attention to the histories of managers. For example, in the case of the managing director of Foods, his initial class location and associated experience as an apprentice must be seen as material contributors to his management practices.

Fourthly, the need to manufacture consent should not be exaggerated. The quality circle programs revealed a reservoir of co-operative predispositions. To argue this is *not* to champion the Mayoite argument that workers are fundamentally driven by some innate desire to co-operate; rather, it involves asserting that managers can often build upon a predisposition to co-operate that is constructed prior to employment and/or outside the workplace (e.g. see Fox, 1971). As noted previously (Littler and Salaman, 1982; Dunford, 1983; Boreham et al., 1986), labour process theory has sometimes been guilty of a 'closed systems' focus in which control is an issue to be determined by intra-organisational action rather than viewing the organisation as merely one (albeit important) arena within which dispositions are formed towards such matters as the structure of authority in the workplace. This predisposition seems to most clearly emerge in response to acknowledgement by management of the skills and knowledge of the workers. In this situation, workers seem particularly responsive (susceptible?) to consensus ideology (see also Knights and Collinson, 1985: 219). Such ideology is not necessarily false for, as Kelly (1985) and Manwaring and Wood (1985) argue, changes in the labour process may include those which benefit both capital and labour. A further point of interest with regard to consent is the possible significance of gender. Given the relative lack of influence by women in unions (e.g. see Ryan and Prendergast, 1982), quality circles may provide a form of workplace participation in which they feel less marginalised. However, there was insufficient attention to gender in the study to allow an informed conclusion on this matter.

Fifthly, the experience of quality circles in the two divisions should lead one to be rather cautious in respect of attributing a fixed meaning to the introduction of quality circles, whether that be the assertion that they are fundamentally mechanisms for control or the notion that they represent some innate harmony of interests. As Kelly (1985) has noted, much labour process theory has incorporated the assumption that redesign of the labour process has a given meaning, an essence (usually increased control), which is independent of context. This essentialism simplified the bases for management action whilst also underestimating the extent to which outcomes are the result of contestation. Associated with this essentialism has been the assumption that reorganisation of the labour process involves deskilling (along Taylorist lines), yet this study clearly shows that it may involve recollectivisation and recognition of the significance of tacit skills (see also Littler, 1982; Manwaring and Wood, 1985; Grzyb, 1984).

The incorporation of these ideas into labour process theory makes it more subtle and less moribund but also less certain in terms of its parameters. Storey's comment (1985a: 207) that labour process theory is emerging as another arena in which sociologists are forced to grapple with the tension in the social sciences between agency and structure is highlighted by interpretation of the actions of the managing director of the Foods division. On the one hand we have a concept of manager as a 'bearer of structure', in this case some notion of the interests of capital; on the other we have clear material conditions (his class and work background) which provide a reading of his actions that does not simply fit with the structural concept of a manager. The difficulty for those interested in working with labour process theory is not to lapse into the determinism that lies behind the attractive closure of early labour process writings nor to produce, albeit unintentionally, a pluralistic voluntarism.

7 The clothing and boot industries, 1880–1939

RAELENE FRANCES

The central focus of this study of the boot and clothing trades in Victoria is the question of worker resistance. As such, it is firmly within the tradition established by Zimbalist (1979), Stark (1980) and Littler (1982) which seeks to balance Braverman's rather one-sided account of the transformation of the labour process under capitalism. Where Braverman dealt almost exclusively with the objective aspects of class, that is, 'class in itself', this later tradition develops the more subjective side of class, 'class for itself'. It is therefore concerned with the question of agency, seeking to determine how far workers were responsible for effecting changes in work organisation and for changing technologies.

Related to the question of workers' responses to changing labour processes is the question of ideology. If, as this study shows, workers sometimes encouraged new methods, such as Taylorism, how far can their response be accounted for by concepts such as hegemony? As Burawoy (1985) argues, were more coercive methods of controlling the workforce being replaced by ideological persuasion?

The third major area of concern in this study is with the consequences of the sexual division of labour for changing work processes. This analysis has two strands. Firstly, there is the question of how far gender influences the social construction of skill. As Beechey (1982) has pointed out, the definition of certain jobs as 'skilled' and others as 'unskilled' is gender-biased. This complicates any attempt to assess the extent to which jobs were 'deskilled' in the Bravermanian sense.

Secondly, there is the question of the relationship between the sexual division of labour in the home and the organisation of manufacturing production. This is an aspect of the labour process debate which has been neglected until recent years, particularly by historians. This study will therefore look at the ways in which the different roles assigned to men and women in capitalist society affected their responses to changing work processes. Clearly, the question of gender is related to the more general issue of agency discussed above. It is also related to ideological issues, in that notions about appropriate gender roles influenced the actions of both workers and employers.

Although these are the major areas of the Braverman and post-Braverman debate which will be taken up here, they are by no means the only important ones. Because of the complexity of the issues, other factors are inextricably linked to the general discussions of agency and gender. The nature of control and the role of technology and the State are other areas which are also explored.

The study

The following case studies concern two of Victoria's major consumer industries in the period from the goldrushes of the 1850s to the Second World War. These years saw the establishment of these trades as highly protected import-replacing industries. The historical analysis is based upon a range of archival material in particular from the Archives of Business and Labour at the Australian National University (ANUA) and Melbourne University Archives (MUA).

The boot industry

As the first Victorian contingent set forth for the Boer War in 1899, it was perhaps only fitting that they were shod in Bedggood boots. Bedggoods was the first boot factory in Melbourne, built in 1854, and for almost half a century had been the leading Victorian manufacturer of footwear (*Australian Leather Journal* [ALJ] 15 February 1900: 836). The boots the soldiers wore to South Africa were similar to the earliest products of the Richmond factory: strong boots, hand-sewn for hard work and long wearing. But the crowds which cheered the colonial adventurers wore a very different article. The civilian boot turned out of the major Melbourne factories in the 1890s was the product of the 'march of the machine' (*People and Collectivist*, 16 April 1898), which wrought a radical transformation in the nature of bootmaking. As the *Australian Leather Journal* noted in 1899, 'there is scarcely an operation which was done long ago with pleasure to the doer which is not now performed by the iron fingers of the machine' (15 December 1899: 753).

The widespread introduction of specialised machinery into

Victorian factories in the early 1890s imposed regular hours, set tasks and other kinds of factory discipline on a group of workers who possessed recognised skills and a correspondingly high degree of independence. 'Chummy' Fleming, one of these 'workers of the old times style', recalled the 'devil-may-care' lives led by the 'Knights of St Crispin' in the 1880s.

> We seldom worked on Mondays. We worked as we pleased; no one feared the boss, for we lived free lives. I remember men at a factory in Fitzroy had been drinking for two days. The boss went to the hotel and shouted for them and they promised to go to work the next day. (*Unity*, 14 March 1938: 10)

Bootmakers were also notorious for honouring 'Saint Monday'. As *The Age* newspaper commented, 'for some occult reason most of the general hands connected with boot and shoe making will not work on the second day of the week' (28 November 1884).

This independence evaporated in the 1890s under the combined effects of economic depression, mechanisation, subdivision of labour processes and the substitution of weekly wages for piece work. In sharp contrast to the carefree ways of the 1880s, workers at Whybrow's factory in 1895 were discharged if they could not complete the task set by Mr Whybrow (minutes, Victorian Operative Bootmakers' Union [VOBU] 3 October 1895; 4 October 1895; 7 October 1895 ANUA). Perhaps even more significant was the report to the Bootmakers' Union in the same year that 'Mr Whybrow still continued to discharge men if they were caught speaking to their shopmates' (VOBU, 18 November 1895). This tightening of factory discipline was symptomatic of the general erosion of job control experienced by bootmakers in the nineteenth century. The erosion was certainly accelerated in the economic storms of the 1890s, but the process had been under way since the middle of the century. Even the largely handmade boot made in Bedggood's first factory passed through a number of different pairs of hands, so that subdivision of the shoemaker's craft was part of factory production from the very beginning of the industry in Victoria.

The labour process in the making of a shoe involved three steps: the cutting out of the leather, the shaping and joining of these pieces and, finally, the trimming and final preparation of the finished shoe (including buffing, colouring and coating the surface of the leather). In the hundreds of 'bespoke' (or order) bootmakers' shops which sprang up in Melbourne in the 1850s, these processes were performed by hand, often by the same person. The handmaker was frequently helped by members of his own family or apprentices. Where women were employed, they usually did the sewing of the uppers, particularly of women's shoes. Apprentices helped out with the 'uncommonly *fidgety* work' of sewing round (*ALJ*, 15 September 1891: 129).

There were over 300 such workshops in Melbourne in 1861, employing between two and twenty workers each, and catering mainly for the well-to-do market. With the imposition of a tariff on imported footwear in 1866, the bespoke workshops were increasingly replaced by primitive factories. In these factories, machinery was used to roll the leather prior to cutting, but the cutting was still done by hand. The actual making and finishing was given to workers who did the work in their own homes (*ALJ*, 16 October 1899: 627).

The introduction of treadle sewing machines in the 1860s drew women into the factories to sew the uppers, although machining was also done in the workers' homes. With the use of the Blake sole-sewing machine from 1876, the male clickers, stuff-cutters, sole-cutters and the female upper machinists were joined inside the factory by increasing numbers of makers, who shaped the upper and fastened it to the sole. There were still no machines, however, for finishing, and the greater output of the factories meant an increasing number of finishers were employed outside the factories. These outworkers were supplied by an army of lads who swarmed over Melbourne, Fitzroy and Collingwood, pushing barrow-loads of boots in various stages of construction (Rimmer, 1974: 115). Nor were the manufacturers anxious to bring the finishers inside the factories, as they took up valuable space and were considered a 'nuisance, trouble and expense as they must have gas burning' to keep their pots of pitch warm (1890 Sweating Report: 8 ANUA).

Despite the advances of the late 1870s, the technology in Victorian boot factories still lagged behind that in England. Power-driven machinery was unusual and the capital outlay on new factories minimal. It was still possible in the 1870s for an enterprising bootmaker with a few hundred pounds to set up as a manufacturer and, with the increased tariff on imported footwear in 1879, more and more bootmakers were encouraged to do so (Linge, 1979).

But the 1880s were not the years of unlimited expansion many anticipated. The increased capacity of Victoria's boot factories was not matched by an expansion of markets. In fact, in the 1880s, imports of footwear increased whilst exports decreased. The result was fierce competition amongst local manufacturers for the remaining markets, and a quest for ways to cut prices and production costs (Linge, 1979: 253–5). In this highly competitive climate, manufacturers sought to reduce their labour costs by intensifying the work inside factories with the installation of gas engines to drive the machines. By 1890, 30 per cent of registered boot factories used motive power, compared with only 12.4 per cent in 1880 (Davison, 1978: 47–8). At the same time, manufacturers attempted to reduce the costs of their hand-labour—the large numbers of finishers working outside the factories. The method used here was much cruder: they simply cut the piece rates. The stage was set for a major conflict between manufacturers and their male workers. The Bootmakers' Union had been growing in strength since 1879 and its members were determined to resist any moves to reduce their wages. Their struggles shaped the changes in methods of production which occurred during the late 1880s and 1890s. We can definitely see in this process the 'march of the machine', as contemporaries were so fond of calling it, but the machines marched to the tune of men, even though the men were sometimes unconscious of the direction they were taking.

In 1884, for example, all of Melbourne's major boot factories were involved in a strike and lockout because the union demanded an end to the putting-out system used in finishing boots. The unionists could see that it was impossible to prevent reductions in piece rates so long as the men were scattered throughout the inner-city suburbs, doing the work in their own homes. By bringing the finishers inside the factories they hoped to organise more effective resistance. The employers did not want this to happen for several reasons. They could see that more effective unionisation would be the result, and they were also unwilling to spend money on buildings to provide the additional accommodation. After a bitter struggle lasting thirteen weeks, the manufacturers agreed to provide this accommodation and bring all the finishers inside (Serle, 1971: 108; Coghlan, 1918: 1486–7).

It seemed that the union had won, but the victory had unexpected long-term effects. No longer able to cut costs by simply cutting wages, employers began to look more closely at other ways of reducing their production costs. Many began to experiment with new machinery and subdivision of labour, hoping simultaneously to reduce their labour costs and weaken the union by replacing hand-labour with machinery (*ALJ*, 15 June 1898: 26). As unemployment rose in the boot trade in the late 1880s, these strategies began to pay off. By the time the depression deepened in 1891, the employers were ready for a full-scale reorganisation of the industry. The first step was to break the power of the union, which they eventually achieved, although it took four years of strikes, lockouts and victimisation of union members (VOBU, 1891–1895). By 1895 the union was left with less than half its 1890 membership and only a trickle of funds as many of the remaining members were on short time. Capitalising on their weakness, the major boot manufacturers installed the latest machines which displaced hundreds of hand finishers. Even worse than the machines from the workers' point of view were the innovations being attempted with hand labour: employers were trying to use weekly wages in conjunction with subdivision to cheapen production on the remaining hand work (VOBU, 13 February 1894).

The bootmakers, raised on piece-work and its relative independence, had in general a horror of weekly wages. Their stated objection was that it usually meant working to a task, or quota, and was thus a concealed form of sweating by speeding-up the workers (*Victorian Parliamentary Papers* [VPP], 1902/03, No. 31, Vol. 2: 347–8). Less openly stated, but just as real, was the fear of factory discipline this implied—a discipline which bootmakers had to a large extent evaded despite the erection of large factories and the installation of machinery. In the Victorian context, weekly wages were invariably accompanied by subdivided labour, a phenomenon the workers were also anxious to resist as it too eroded their control over their work and devalued their skills. So it is not surprising that when the Employers' Association suggested subdivision of hand finishing in 1894 the workers were unco-operative. They were prepared to accept it in conjunction with machines, but not on hand labour, even if the employers guaranteed to maintain their current rates of pay (VOBU 13, 22 February 1894). But their opposition was futile, as they had neither the funds nor the solidarity to wage a successful strike. By 1896 all the leading firms had introduced the team system, where necessary discharging recalcitrant or slow workers and replacing them with faster, more docile men (VOBU, July–October 1895; VPP, 1902–3, Vol. 2, No. 31: 440). Some employers imported English operators who were already used to the new methods. Those workers who refused to work under the new regime were forced to finance their independence with the meagre earnings they could get repairing machine-made boots and shoes.

Although this story has such an unhappy ending (at least from the workers' point of view), it nonetheless contradicts the theory that certain forms of work organisation and certain technologies are the result of unimpeded capitalist initiative. Had the workers not succeeded in the 1880s in eliminating outwork by finishers, the employers would have responded quite differently to the economic crises of the 1890s. Faced with dwindling profits, it would have been far simpler to cut the costs of their labour by reducing wage rates. Where outworkers were the norm this was quite an easy matter. Besides, had finishing continued as an outwork process, it is less likely that manufacturers would have considered putting up new machinery as this would also have meant extending their buildings.

Had it not been for the 1884 lockout, the boot industry in the 1890s would most likely have adopted the same course of action as that followed in the clothing trade: extension of the outwork system on ever-decreasing piece rates. As the clothing trade provides a useful comparison to the case of the boot industry it is to that which I now turn.

The clothing industry

Just as the actions of workers in the boot trade influenced changes in the work process, so too did workers in the clothing industries affect the organisation of work. Here too the question of outwork was crucial. The first Clothing Trades Wages Board Determination in 1897 virtually put an end to outwork in the making of men's and boys' clothing. By setting a piece rate on a slightly higher scale than the weekly wage, manufacturers were discouraged from using what had previously been cheaper outwork labour. (Outworkers were legally required to be paid by the piece.) Denied this source of cheap labour, they adopted the same course as the boot manufacturers: they installed power-driven machinery and put their machinists on a weekly wage with a set task. By thus speeding up the workers they were able to keep labour costs to a minimum (Frances, 1986: 98). The important point here is that it was the workers, through their representatives on the Wages Board, who secured the higher piece rates and thus indirectly precipitated the change to full-scale factory production in the twentieth century. From the beginning, in 1898, the tailoring industry in the years up to the Second World War continued to modify its methods of production. The use of power-driven machines was soon followed by increased subdivision of the making of garments, particularly trousers and vests, and this process was accelerated with the introduction of new patent machines to perform specialised operations such as button-holing, button-sewing, hemming and tacking. One woman who worked in a clothing factory under the new regimes described the experience:

> The roar of the machines filling the air until it hurts, deafening the ears and pressing on the brain like a tangible weight; the monotony of the work, and constant effort not to fall behind with it; the endurance, both mental and physical that must be sustained, impose a strain on the nerves none but those set down among the conditions can realise. (*New Idea*, 6 July 1903: 45)

Although this description of the nature of work under the new methods sounds particularly unattractive, the workers in the industry initially encouraged the changes. But their attitudes changed with time. For the most part, the official attitude of the Clothing Trades Union in the period up to the 1920s was one of encouragement of the new methods as a way of improving the efficiency of the industry in general. The logic was that increased productivity would mean a greater capacity on the part of the industry to pay higher wages, reduce hours and improve factory conditions. At this stage it was more a question of collaboration than resistance. Any resistance that did exist was to the debasement of skill, but this was usually restricted to insistence that apprentices initially be taught all aspects of their craft, even if thereafter they were only required to use part of this skill. There were also fears that the new machines would throw workers out of work, but these were shortlived when it became clear that they were unfounded.

Rather than resistance, the union's attitude during this period could be more accurately described as active encouragement of the new methods. The union's support for the new methods reached its peak just after the First World War and was clearly articulated in the first case before the Federal Arbitration Court in 1919. Not only did it claim reduced hours and increased wages on the grounds that workers should share in the benefits of increased productivity resulting from the new techniques, but it also argued that such concessions would force employers to be even more 'efficient'. Alf Wallis, union organiser, spoke of the 44-hour week, encouraging employers to improve their efficiency by adopting 'good methods'. There was no doubt what he meant by 'good methods' as he referred to Taylor's works on the subject of the division of labour as a guide for those employers who needed instruction (1919 Arbitration Hearing: 295–303 ANUA).

Nevertheless, it was already apparent that workers were not entirely happy with the changes in their working lives. In the same claim, the union called evidence to show that reduced hours were justified for another reason—that the new methods caused additional stress to the workers. This was the argument which ultimately convinced Higgins, who said that he could 'not ignore the evidence as to vibration, as to the artificial light, as to the strain from speed, as to the hurry, the noise, the fluffy particles, as to the long sitting and stooping, the fatigue, the frequent headaches and injury to eyesight', all of which were worse under the team system of production (Higgins' judgment, *Archer* case 1919: 14 ANUA).

The early 1920s saw an increasing conviction within the union that the workers were becoming 'the victims of the Juggernaut of industry', to use Higgins' colourful phrase (Higgins' judgment, *Archer* case, 1919: 3). From its aggressive stance in 1919, demanding a share in the higher profits, the union retreated somewhat in the face of greater experience with the new methods. It had become clear that the employers not only wished to cheapen production by increasing output through subdivision, but that they desired to reduce costs even further by eliminating margins for skill and speeding up the workers. Instead of talking about workers sharing in increased profits, the union's emphasis shifted to insisting that it was not fair that employers should reap greater profits at the expense of the workers.

This disillusionment with the employers' sense of justice was even more evident in the 1927 Arbitration Hearing. It was also obvious that more workers were dissatisfied with the effects of the new methods. They complained about the bad effects the noise, speed and monotony had on their health. The Pelaco factory, model of the new industrialism, was singled out for attack. Rather than being the 'Home on the Hill' it claimed to be, evidence was produced to show that half the employees left before giving three months' service and 80 per cent left within twelve months. Of those who stayed, over 60 per cent were under 21 years of age (1927 Arbitration Hearing: 4144 ANUA). Even for an industry with a high proportion of juniors, this was extraordinary.

Despite these murmurings of discontent, it took the economic depression to raise serious doubts about the whole system. The popular belief among clothing workers that the depression was caused by over-production caused doubts about the advisability of endlessly increasing output, whereas the increase in the number of juveniles working in the factories prompted further concern about the effects of deskilling on the workforce. These doubts and fears did not, however, result in a change in union policy, and the officials still pursued their limited aims of 'preventing the worst features of the new methods from being applied and raise [sic] the price of labour' (Clothing Trade Union File, 'Strikes, 1916–1935', MUA).

This view was not shared by all the union membership. Women machinists at Melbourne's three largest factories, Ellinson's, Sackville's and Frieze Brothers', announced their intention to strike in March 1935, while the union secretary (Alf Wallis) was in Tasmania. They were refusing to operate under a new team system which further subdivided the making of trousers. The union officials rushed in to negotiate on behalf of the women and the strikers were successful in temporarily modifying the new system and also in securing an increase in the piece rates. It was quite obvious that the union leadership had been caught off balance by the action of the women and Wallis justified his lack of enthusiasm by saying that:

> in every industry, everywhere, no union or strike has been able to prevent new methods and machines . . . and . . . that if employees struck against acceptance of the new machines they would not have one chance in a hundred of preventing new methods. (CTU file, 'Strikes, 1916–35')

He neglected to mention that the Clothing Union had helped establish these methods and continued to restrict its real activity to the amelioration of wages and hours rather than attempting any fundamental reappraisal of production methods.

There were a number of reasons for this union's attitudes towards mass production, subdivision and mechanisation. In the period from the 1890s to the 1930s, there were particularly strong pressures to co-operate with management. The union itself had its origins in the exclusive Tailors' Protection Society which actively collaborated with clothing manufacturers in the nineteenth century to achieve a high level of tariff protection for the industry. This was seen to be of benefit to both masters and men as it was supposed to guarantee the viability of the industry as a whole. It created a strong sense of mutuality of interest which weathered the 1890s depression and influenced official union decisions at least until the Second World War. In arguing for higher tariffs, unionists were brought face to face with questions of productivity and competitiveness which made them far more sympathetic to employer moves to 'rationalise' production than would otherwise have been the case. As late as 1921, the federal secretary, Herbert Carter, explained the union's reluctance to push for equal pay for women on the grounds that the employers could not stand the additional wage bill without prohibitive tariff increases (Frances, 1986: 104).

In addition to the fiscal question was the problem of sweating, which was also a recurrent political isue throughout this period. In this case the alliance was between the 'reputable' manufacturers and 'skilled' workers, who combined to prevent the undercutting of wage levels by so-called 'sweating employers' and non-union labour. The employers who paid acceptable wages in the 1890s were usually the larger manufacturers who had regular contracts with the Flinders Lane wholesalers. The 'sweaters' had less stable demand for their products and lower profit margins, and consequently relied on keeping the price of labour at a minimum to enable them to compete in an oversupplied market. From the union's point of view, larger factories usually meant more efficient production techniques and fairer wages so they were prepared to encourage centralisation of capital and to ally with the large manufacturers against the smaller firms. Again, Herbert Carter was speaking in his official capacity as federal secretary when he said that 'it suits us to have the industry in more or less the control of the bigger people' (1927 Arbitration Hearing: 1561). This acceptance of large scale manufacture implied an acceptance of mass production techniques, mechanisation and subdivision. It also fostered a trust between the officials and the employers. Alf Wallis, the Victorian branch secretary, testified before the 1927 Arbitration Hearing regarding the method of fixing piece rates. He explained that 'we get assistance from other employers who are in a better position to fix the price than the worker' (1927 Arbitration Hearing: 1731).

A third factor which encouraged co-operation between the union and the employers was the frequency with which 'skilled' workers became proprietors of their own factories and workrooms. This was easier in the clothing field than in many industries, as the initial capital outlay was small and sewing machines were readily available on time payment. Not only rank and file union members but also officials left the union to become employers. The prospect of becoming a capitalist affected individuals' attitudes to management perspectives and, as late as 1944, the Clothing Trades Union was described by a female worker as being 'hand in hand with management' (Roberts, 1980: 412).

A further factor influencing the attitude of the union to capital was the male-dominated nature of the union and the sexual division of labour within the industry. Males were concentrated in those areas traditionally seen as skilled occupations, namely high quality tailoring, cutting and pressing. They were also largely confined to the men's and boys' clothing and women's tailoring sections. Women predominated in the dressmaking, millinery, shirts and underclothing sections, and in the machining of trousers and vests. And, although women formed approximately 85 per cent of the membership of the Clothing Trades Union in the twentieth century, the executive committee was structured so as to give dominance to the men (Frances, 1986: 106). The positions of president, secretary and treasurer were never held by women, and only in rare circumstances were there paid women organisers. Women were never chosen as delegates to the federal council, and only occasionally did they represent the union on the Trades Hall council.

The sexual division of labour within the workplace was thus reflected in the union structure. Men held those positions with the highest status and power, while women formed two subordinate levels: a layer of assistants and, below them, the mass of low-status, poorly paid rank and file membership. The result is not surprising. The male elite who were the union stalwarts and officials identified more closely with management than with the ordinary female machinist. They were able to bargain with employers to preserve their own margins for skill at a relatively high level while at the same time accepting the nominal margins awarded to women. When male tailors, pressers and cutters received an 18 shilling per week margin on the base rate, most females received from one shilling and threepence to three shillings per week for their skill. In referring to the origins of these margins in Wages Board Determinations, the Victorian secretary explained that:

> To my own definite knowledge I know that male representatives on these tribunals deliberately refrained from using their best endeavours to improve the lot of the women mainly on account of sex jealousy in relation to industry. At one time I heard a representative say 'Oh let the women look after themselves. They have taken our jobs . . . ' (H. Carter, 1922 Arbitration Hearing: 22 ANUA)

This bargain also suited the employers, since they could well afford to pay the 18 shilling margin to 10 per cent of their workers when the rest were only paid a few shillings each.

But what of the mass of women operatives? As a group, they were far more diverse than the male clothing workers, since they were employed in significant numbers throughout every branch of the trade. Large numbers were also employed outside the registered factories, working in small workrooms or in their own homes. It is therefore very difficult to generalise about their attitudes and responses. But, despite this diversity, it is possible to differentiate two broad groups of women workers. On the one hand were women employed in the more organised and traditionally male sections of order and ready-made tailoring. On the other hand were the majority of women, who were engaged in dressmaking, millinery, underclothing and whitework. This group was twice as numerous as the tailoresses.

The tailoresses had a tradition of unionism going back to the nineteenth century. Their first strike in 1882 produced short-term gains in the pay and conditions of employees in shops and factories, but these gains were significantly eroded in the late 1880s and 1890s as an increasing amount of work was contracted outside the factories at rates well below the union's log (Fry, 1956, passim; Brooks, 1983: 36–7). It was only after the first Wages Board Determination in 1897 fixed piece rates for both factory employees and outworkers that it became uneconomical to employ outworkers on men's and boys' clothing. With most of the work being done in factories, the Tailoresses' Union was able to organise more effectively and was firmly re-established by the time of its amalgamation with the male unions in 1907. Tailoresses continued to play an important, if secondary role on the union executive throughout the years to the Second World War. There appeared to be a general acceptance of the union and support for union principles.

Nevertheless, the continued dominance of official positions by men and their hostility to women organisers created a certain amount of suspicion amongst the rank and file membership. There was a widespread feeling that all the union officials cared about was collecting members' subscriptions. In 1938 a woman working at Sackville's factory expressed this feeling in a letter to the secretary, saying: 'The girls at the factory say all the union worries about is that you pay Mr Wilson your contributions . . . ' (Correspondence CTU, 9 November 1938 MUA). This scepticism about the officials did not prevent the women employing union tactics to achieve their own goals. There are dozens of incidents of groups of tailoresses striking for improved conditions or increased wages. These episodes occurred throughout the entire period, but were particularly frequent during times of labour shortage in the tailoring section. Tailoresses were fortunate, as they had been able to restrict the proportion of untrained workers to journeywomen. (The threat tailoresses posed to tailors had ensured the tailors' support over this issue.) Relative to other women clothing workers their market situation was therefore quite favourable.

Their strategic importance was also sometimes recognised by the union officials, who in general despaired of effectively organising women. As a Victorian organiser told the 1937 federal council, the union 'had portions of an army however and could carry on a guerilla warfare' (minutes, Federal Council Meeting, May 1937 MUA). In this instance he was referring particularly to the strike of Melbourne trouser machinists referred to above.

Although by the 1930s tailoresses were more usually called machinists, it is interesting to note the continuing influence of craft ideas. In the nineteenth century the Tailoresses' Union was careful to maintain strict divisions between stock and order hands, and refused outright to have anything to do with outworkers on the grounds that such an association would 'drag them down' to the outworkers' level (Frances, 1986: 108). The resistance which groups of tailoresses showed in the twentieth century to the increasing subdivision of their work also bears some relation to their reluctance to see their skill debased. The new techniques had more important effects on the working lives of women in tailoring, since the men were often compensated for loss of craft skills by being made supervisors. For the women there was no compensation, only increased speed, less job satisfaction and less flexibility in seeking employment. Throughout the period from the First World War until the late 1930s, tailoresses were able to prevent the wholesale subdivision of the processes of making men's clothing. Even when they did concede some changes, they insisted on agreements to maintain wage rates and to ensure that apprentices were trained on all sections of a garment, even though they were later restricted to making sections of the garment only. In fact, the increase in larger factories and the demise of the small order tailor improved the ability of tailoresses to organise effectively, since it broke down the barriers between order and stock workers.

In contrast to the tailoresses, women working in the dress, shirt and underclothing sections had no traditions of craft unionism. Unions formed around the turn of the century to organise for Wages Board representation were industrial rather than craft based, and included outworkers as well as factory employees. However, despite the efforts of a small group of women activists, these unions failed to achieve mass support and were financially ill-equipped to achieve any progress in that direction. The difficulties in organising these sections were truly daunting. About one-quarter of the employees were outworkers and, when outwork in the men's clothing section virtually disappeared after 1898, the competition amongst outworkers in these female sections became even more intense. Furthermore, the Wages Boards refused to fix piece rates in these sections on the ground that the diversity of operations would make it too difficult to set individual rates. Instead, the board left the fixing of prices to employers, who were supposed to set the piece work price at a level to enable an average worker to earn the minimum weekly wage. No guidance was given as to what constituted an 'average' worker, so that employers were able to keep prices at sweated rates with impunity.

From the early twentieth century, the situation for the workers changed significantly. Larger factories drew more juveniles into factory work, although outwork still continued. The proportion of males to females engaged in these sections also altered dramatically over the period from 1901 to 1933, as women who were self-employed and small employers were replaced by male proprietors and managers in large factories.

For the outworkers, mass production and specialisation in factories meant more pressure to cheapen their labour to compete. Outworkers were isolated from any union organisation, especially after the amalgamation of the women's unions with the Federated Clothing Trades Union in 1915. The male officialdom failed to appreciate the advantages that outwork offered to women of all ages, who were able to combine their domestic duties with wage-earning, however poorly paid the work was. This was true for both mature women and young girls who had to contribute to the support and care of parents, husbands, brothers, sisters and children. Like the bootmakers before them, the official union attitude opposed outwork in principle on the ground that it was impossible to regulate, but this position offered no solution to the women who faced the double task of earning money and caring for families. The failure of the union to abolish outwork reflects this fact.

Those women who did work in factories were potentially easier to organise. However, the employers succeeded in preventing attempts to limit the numbers of juveniles entering the trade, so that in effect the industry was always oversupplied. When in 1919 the employers appealed to the Court of Industrial Appeals against a decision of the Underclothing Wages Board to limit the proportion of juveniles to one to every two adults, they based their case on the high wastage of female workers. This was attributed to the pressures on females to leave factory work to take up home duties of various sorts. Their appeal succeeded in altering the proportion to three juveniles to every two adults. Unlike the tailoresses, they could not argue that fewer female juveniles would mean more work for adult males.

Not content with this advantage in obtaining labour, employers adopted other methods to ensure that this section of the clothing industry remained non-unionised. Unlike the union, certain employers were prepared to make allowances for women's unpaid responsibilities and to use this to their advantage. One such employer was the Pelaco shirt company. James Law, managing director and principal shareholder of the company, explained how his system of manufacture, based on Taylorist principles, also suited the needs of the labour market. He maintained that in most cases industrial employment was seen by girls as a stepping stone to matrimony and that the 'present-day girl . . . is looking for work where she can get the most money in the quickest time'. Furthermore, he said, 'the man who is able to put his industry into such a position as will cope with that demand by the girls is the one, I think, who is best able to achieve satisfactory results' (1927 Arbitration Hearing: 4125 ANUA). Of course, Law was only talking about a section of the workforce, not the large numbers of married women who later returned to the paid workforce.

Law's method of manufacture involved minutely subdivided tasks and high speed production which few workers could sustain for long periods of time—hence the extremely high labour turnover. His use of young girls, who in any case expected only intermittent employment, ensured that dissatisfaction was expressed in this high turnover rather than in opposition to his methods. Any possibility of more organised dissatisfaction was averted by his paternalistic industrial relations style: he provided dining rooms, tea-breaks, social clubs and even a factory newspaper through which he disseminated his philosophy of co-operation between capital and labour. Numerous attempts by the union to organise Pelaco workers met with dismal failure.

Conclusion

What the case of the clothing trades demonstrates is that the question of worker resistance is very complex and closely connected to the question of gender. Different perceptions of the work process, based on different experiences of work, created divisions between male union officials and the female membership and prevented a united stand to preserve job control. Because the union officials were so close to management they were susceptible to ideological persuasion from theorists such as Taylor. The female workers, for whom experience on the shop floor acted as a powerful corrective to theory, were more inclined to resist innovations along scientific management lines. But, for them, the sexual division of labour, both in the home and the factory, was a major obstacle to organisation. Only the tailoresses, who comprised significant numbers of mature young women with time and energy to organise, could offer any effective resistance. They were also in a favourable labour market position and drew on a strong tradition of craft unionism. The majority of female clothing workers, who were employed in dressmaking and whitework, were, by contrast, either very young or burdened by domestic responsibilities. Faced with an oversupplied market, high labour turnover and extensive outwork, organised resistance was unlikely. Capitalists were also able to use the role expectancy of juvenile female workers inside factories to encourage passivity and co-operation in implementing scientific management techniques, whilst at the same time exploiting divisions between male union officials and the female workforce.

The sexual division of labour in the family and in the workforce was thus reinforced by the activities of male unionists, employers and the arbitration system. Although this resulted in gains for the 10 per cent of clothing workers who were male, these gains were at the expense of the poorly paid and increasingly overworked female workers. Ultimately, it was also at the expense of a united workers' resistance to the erosion of their skill and the intensification and fragmentation of the work process.

The bootmakers were not divided in the same way by sex. But the question of gender nevertheless played an important part. The

Bootmakers' Union succeeded in curtailing outwork partly because the workers were men. Unlike female clothing outworkers, they were not usually tied to the home by their domestic duties within the family. The sexual division of labour in the home in fact encouraged the absence of the male 'breadwinner', making the men's move to the factory both possible and acceptable. The dirt and noxious fumes generated by boot finishing in fact made the removal of these processes from the workers' homes highly desirable, making the housewife's task easier and the general environment healthier. And whereas this was also true of clothing outwork, the male bootmakers could contemplate leaving the home without having to balance responsibilites for childcare and nursing with the demands of wage-earning.

Male boot workers thus had the option to trade the independence of the outworker in order to maintain their earnings. But their decision in 1884 had much greater long-term effects than anticipated. The bootmakers hoped to resist factory discipline and the degradation of their skills by opposing weekly wages and subdivision, but their earlier action in encouraging the extension of the factory system was to place them at the mercy of the employers and their machines. The other concession they made in agreeing to the use of weekly wages and subdivision in conjunction with machinery was also counterproductive, as it too encouraged the installation of even more machines and opened the way for less skilled labour, mainly boys, to be employed in place of tradesmen. And whereas Ramsay (1985) has noted that capitalist strategies to change the labour process have on occasions backfired, the case of the Victorian boot industry shows that workers' resistance to these 'advances' can also backfire.

The boot and clothing cases thus contribute to a fuller understanding of the factors involved in changing work processes than that offered by Braverman. Clearly workers as well as capitalists were agents of this change. But close historical analysis reveals that it is invalid to conceptualise the role of workers simply as impediments to capitalist innovation. Workers in both industries did resist changes at certain times, but they also contributed to innovations by their strategies to improve their general position. Conversely, employers did not always seek to extend their direct control over the work process. As both cases show, manufacturers sometimes saw the massing of workers under factory roofs as less expedient than keeping them more independent but less organised as outworkers. The logic of capitalism did not demand scientific management in these formative years of Victorian manufacturing any more than it does today.

8 Technological innovation in health care

JEANNE DALY AND EVAN WILLIS

Most research in the labour process tradition has involved examining the organisation and control of the labour process in work contexts such as factories and offices where the *direct* capitalist relations of domination and subordination involved in wage labour exist. In this chapter, by contrast, the questions which Braverman and subsequent analysts posed about the organisation and control of the labour process are examined in the context of *indirect* wage labour relations in that area of the work force known as the professions. Drawing upon earlier work (Daly, 1985; Daly and Willis, 1987), we take as our case study the labour process surrounding technological innovation in the field of medical diagnostic imaging.

As has been argued in the introduction to this book, one of the major lines of critique of Braverman's work has been that some concepts which are essential to an adequate analysis of the labour process in advanced capitalist societies such as Australia are either missing from or insufficiently developed in his analysis. In this chapter we argue, following Burawoy (1985), that two essential conceptual ingredients, ideology and the capitalist state, are largely undeveloped in Braverman's work but are important to an adequate understanding of the professional labour process in general and technological innovation in particular. 'Ideology' is used here to mean a complex and contradictory set of ideas, representations, actions and practices grounded in social relations. That is, ideology consists not only of ideas, rationalisations and justifications, but also has a material force in concrete practices (i.e. what people actually *do*).

Our starting point is with class analysis in the world of work, an issue which Braverman brought back to the centre stage of sociological analysis. Although class relations between the dominant or ruling class and the working class in advanced capitalist society constitutes the major structural relation of domination and subordination shaping these societies through the social division of labour, a comprehensive analysis also requires account of an intermediary class grouping. The middle class, far from withering under the impact of proletarianisation as Marx predicted, has remained and indeed developed as an important structural feature of capitalist societies such as Australia and New Zealand. Our concern is with those occupations recognised as being professions, which constitute an important part of the new middle class. Indeed, professionals are defined as new middle class irrespective of their employment status as wage labourers (i.e. salaried) or self-employed. Following Gramsci (1971: 12) it is on the basis of their contribution to the maintenance of ideological hegemony, the basis of capitalist social order, that they are so defined. Professionals such as members of the medical profession (henceforth doctors) act as organic intellectuals to the dominant class, mediating the relationship of domination and subordination between classes and reproducing labour power by, for instance, writing medical certificates to legitimate work absences or conducting pre-employment screenings to assess fitness to enter a wage labour contract. Their importance at the level of ideology is that they are 'experts in legitimation' (Merrington, 1968: 154) who 'act as the mediators of the realities of capitalism into values' (Davidson, 1968: 45).

The conceptual confusion with the terms 'profession', 'professionalisation' and 'professionalism' has been considerable. Whatever utility the term may have in commonsense language, it has limitations as an analytical category for labour process studies. Clearly the phenomenon must be examined historically. The professions as we know them today, originated with the advent of capitalism in the nineteenth century (Laslett, 1971) and their expansion provided a link between the educational system and the occupational order (Schudson, 1980). Initially the professions were seen positively as a source of stability and integration in a modernising world (Durkheim, 1957). More recently, professionalism has been seen as a claim to a particular form of self governance of the occupation, to maintenance of internal self-control over the occupation (autonomy). Professionalism could thus be analysed as an occupational ideology which legitimates autonomy (see Larsen, 1977). As Freidson argues, professionalism is 'a deliberate rhetoric in the political process of lobbying, public relations and other forms of persuasion to attain full control over its work' (1970: 80).

In labour process terms, then, professionalism can be seen as a form of resistance to the imposition of control external to the occupation itself, a strategy for maintaining the unity of conception and execution in the performance of work. However, it must also be analysed as a strategy involving quite explicit gender and class aspects. Regarding gender, as Versluysen argues, 'theorists of professions ignore the fact that groups which have successfully professionalised have usually been male in composition' (1981: 22). While ideology in this or any other context does not operate in a straightforward, even way, but in a complex, contradictory uneven fashion, in class terms the ideology of professionalism reproduces aspects of the dominant class ideology and promotes hegemony. In particular, professionalism reproduces what Habermas (1970) calls the 'ideology of expertise', which claims that only those who have specialised knowledge can adequately make decisions about appropriate courses of action (such as treatment), despite the ease of demonstrating that, for instance, many medical decisions are made as much on moral and political grounds as they are on purely technical grounds. The claim to skill or expertise, furthermore, is articulated as beneficent power based on the notion of technological rationality, the claim to effectiveness and on individualism. Yet in class terms, as Larsen (1977) has shown for a number of capitalist countries, professionalism legitimates inequality and elitism. Professionalism also has important ideological consequences in blinding the facts of relative subordination and powerlessness by the professionals themselves, thus pacifying them.

Turning now to the professional labour process in the area of health care, two features are immediately apparent. One is its hierarchical nature, the other its domination by one occupation—the medical profession. The labour process is constituted as a very complex division of labour in part negotiated on a day-to-day basis (see Freidson, 1976), but overwhelmingly formally imposed and codified. The process by which this division of labour developed and medical dominance was achieved and defended has been analysed in detail elsewhere (Willis, 1983).

Professional expertise and skill

In this section we consider the issue of skill in the context of the professional labour process, using health care as our example. Our starting point is to note that, in the professional labour process, skill tends to be constructed as expertise. Although the two terms are in many ways synonymous, an aspect of the ideology of professionalism is to construct skilled professional workers as 'experts'. Being 'professional' (rather than amateurish) involves bringing expertise to bear on a particular range of issues. The difference between the conceptions of skill and expertise, insofar as it exists, is in the greater emphasis on knowledge (mental labour) in expertise. A skilled worker expresses a unity of manual and mental labour, of hand, eye and brain. A professional expert, by contrast, is more likely to be concerned with the design or planning than with the execution of a task. The expertise of a doctor is more concerned with defining illness and devising a strategy for treatment than with the execution of the treatment. This gives rise to the phenomenon of 'pass-the-task' whereby, over time, the more mundane, routine, manual tasks are delegated to lower order occupations. Obvious examples are ward-based hospital care by nurses and the dispensing of drugs by pharmacists.

Nor is this process of differentiation restricted only to doctors. It occurs as 'subprofessional dominance' as well, further down the hierarchy. The creation of the occupation of state-enrolled nurses (formerly nurses' aides) is a good example of this process, as the more mundane manual tasks of nursing (such as emptying bedpans) have been delegated to subordinate occupations.

This mental/manual distinction fails to account for the high value attached to the skill of, for example, a cardiac surgeon, and here a useful analytical device is the distinction, made by Jamous and Peloille (1970), between elements of technicality and indeterminacy in the work performed by an occupation. The technical aspects are those which are susceptible to codification by rules, procedures and techniques and can be made available to others in the form of a manual (e.g. how to bandage a sprained ankle). The indeterminate aspects, by contrast, the personal 'rules of thumb' such as interpersonal skills, clinical judgement, 'flair' or 'knack', are difficult to codify and cannot easily be taught to others (Atkinson et al., 1977). All occupations contain a mixture of technicality and indeterminacy, of explicit and implicit expertise, and it is possible to compare occupations in terms of the ratio of indeterminate to technical elements (the I/T ratio). In the health labour process, the medical profession has a high ratio of indeterminate elements, sometimes called 'medical mystique' (Atkinson et al., 1977), to technical elements. It is this claim to uncodifiable indeterminacy in medical expertise which has provided a powerful means of resisting technological deskilling. The more codifiable aspects of clinical practice have, by a process of specialisation of the division of labour, been incorporated into the training of subordinate occupations and it is *their* susceptibility to deskilling which provides a buffer for the medical profession.

What then is skill/expertise in the medical profession? While not denying objective criteria for the concept (notably the existence of job autonomy), for the purpose of this analysis, skill/expertise is best seen as socially constructed, as a form of social status, created and maintained through politically limiting the expertise of other competing occupations by a combination of custom and tradition and collective organisation through trade unions/professional associations. Historically the Australian Medical Association (AMA), in classic trade union fashion, has provided resistance to alternative modes of organising and controlling the labour process in health care.

The AMA strategy for establishing and maintaining dominance and control over the labour process has been two-pronged. At the level of their own work, of *autonomy*, they have attempted to create and defend the direct employment relationship between doctors and their clientele, patients. Two phases of this struggle have been apparent; firstly, to rid themselves of salaried employment with Friendly Societies (see Green and Cromwell, 1984) and, secondly, to resist the entry of government as a third party in the various proposals for salaried service along the lines of the British National Health Service. Uniquely, in Australia, they were able to mount a successful constitutional case against salaried service, succeeding in the High Court in 1949 with the highly questionable argument that salaried medical service constituted a form of civil conscription, something the Australian constitution proscribes. The second prong of the strategy to establish dominance of the organisation and control of the health care labour process is at the level of the work of other health occupations, represented as a relationship of *authority*. Medicine has gained the right to legitimately evaluate and direct the work of many other health occupations, while at the same time denying the legitimacy of the converse. This organisation of the labour process in health care does not occur in a vacuum, however. Medical dominance operates with legal backing to control the labour process. An analysis of the professional labour process therefore requires taking account of the fact that the State provides the medical profession with a legally enforceable monopoly over both its own services and those of others (represented as *medical sovereignty*). Briefly, we are using the concept of the democratic capitalist State as a balance of forces and relations within a material framework and organisation constituted by an array of judicial, legislative, military and coercive institutions. The State serves the interests of the dominant class through maintenance of the economy, political order and legitimacy, without being directly subordinate to (i.e. having relative autonomy from) it.

Medical dominance and control over other occupations are maintained through State patronage of the profession in the form of legislative backing. Statutory registration legislation legally defines the occupational territories or task domains not only of medicine itself but of other health occupations as well. Radiographers, for instance, cannot legally interpret the x-rays they take. The struggle over licensing legislation at the level of the State has been important to the emergence of the labour process.

State patronage for medicine is central to the organisation and control of the labour process. The autonomy, authority and sovereignty of the medical profession depends upon this State patronage. The question, then, is why has the State endorsed this medical monopoly? The answer, argued in detail elsewhere (Willis, 1983) is because of the class compatibility of the paradigm of scientific medicine. In the creation of ideological hegemony, the form of medical knowledge which saw disease as individual and biological rather than mediated in a social and political manner had a definite compatibility with dominant class interests. State patronage for the dominance of medicine has been the result (see also Berliner, 1985). This historical alliance between new-middle-class doctors and the dominant class has been the major factor giving shape to the labour process in health care.

Having considered the organisation and control of the labour process in health in general we now turn to the process of technological innovation in particular. This section analyses the role of imaging technology in some of the changes which have occurred in the health care division of labour over the last century.

The study

This case study of the impact of technological innovation on the professional labour process is based upon content analysis of documentary evidence. The documentary analysis has been both historical and contemporary. The former has involved tracing the emergence of a specific labour process surrounding a central diagnostic technology of medicine (i.e. its production). The latter has focused upon its 'reproduction'; that is, how the labour process surrounding imaging technology has developed, with the latest step in this development being a recent Victorian governmental inquiry. The ostensible reason for this inquiry was a shortage of radiographers in public medical practice, but the content of the report has been analysed as the latest phase in the development of the labour process in this field.

The concept of technology used here includes not only a *material* component, the machine, but also a *social/ideological* component, the social relationships involved in its use. Here the emphasis will largely be on technology used to image the living body, in particular the various versions of the x-ray machine. Analysing the x-ray machine as a technology, we include the machine, the skills involved in its operation and the social relationships which exist between the radiologist, the radiographer, other occupational groups such as chiropractors and the patients in whose interests the technology is used.

This section of the chapter has three parts. In the first we argue that imaging technologies played an important role in establishing and maintaining medical dominance. In the second section we analyse the role of x-ray technology in the labour process which developed following its introduction. Lastly we trace these developments to the relationship between the medical profession and the State, arguing that the technology which helped establish medical dominance may now be providing the State with a justification to intervene in the health arena, This, however, is not to argue that the technology in any simple way determined these changes. Indeed, it is specifically against this notion of technological determinism that we wish to argue. Our argument is for a dialectical relationship between technology and society.

Before analysing in detail the role of imaging technology in medicine it is necessary to look briefly at the more general development of technology in medicine, particularly its role in the nineteenth century when the profession established itself. It is also necessary to draw a distinction between technologies directly involved in the treatment of disease (therapeutic technologies) and our concern here, those involved in the identification of disease (diagnostic technologies).

The rising status of the nineteenth century medical profession rested heavily on the claim to being scientific, and this in turn led to a sometimes indiscriminate pursuit of diagnostic data (Corrigan, 1828: 588). With only a very few technologies capable of providing this data on living patients, the status of the medical profession rested squarely on the extensive use of such technologies as were available. Claims of scientific objectivity have a political function in establishing control over an area of expertise and protecting it from invasion by others.

Ideological arguments used to justify the autonomy of the medical profession mask the importance of that profession to the capitalist state. Diagnostic technology plays a key role here, as is aptly illustrated by the importance attached to the use of the ophthalmoscope in nineteenth century medicine. This instrument was not used primarily because its diagnostic potential was markedly higher than that of direct patient accounts, but because it produced objective evidence of disease, that is, evidence that was independent of the deceit of 'malingerers, conscripts, applicants for pensions, insurance or society benefits, plaintiffs claiming damages for injury . . . ' (Jackson, 1894: 337). But, above all, medical science is an important component of the ideology of scientific progress and technological advance. The benefits of science can only be delivered by the scientific experts who are able to address the complex problems involved. If we want these benefits then we have no option but to 'trust the expert'. The means by which these benefits are made available to consumers is technology, and 'progress' is therefore measured by technological advance. When applied to medicine, this means that the status-enhancing technologies, whether therapeutic or diagnostic, must always be the newest and the most powerful 'weapons' against disease. With new technologies constantly replacing old, technical skill has to be continually renewed and is therefore resistant to codification. Esoteric technical skill, however, is only one aspect of the expertise claimed by professionals. The other aspect of expertise is the claim to indeterminacy, here analysed in the next section where we turn to a case study of the use of x-ray diagnosis in medical care.

X-rays: technology and authority

X-rays were discovered by Rontgen, a physicist, in November 1895 and were a mysterious radiation (hence the name) which penetrated the soft tissue of the body but were absorbed by the more dense tissue so that an image of the internal structure of the body could be captured on a photographic plate or a fluorescent screen placed on the opposite side of the body. By January, 1896, the first x-ray pictures for clinical purposes were taken by a British electrical engineer (Shanks, 1950: 44) and the diagnostic potential of the rays became clear. X-ray machines became readily available and easy to use, and were sold in large numbers. It became a popular parlour game to produce images of the internal structure of the human body. Such parlour games rapidly lost their fascination when, within two years of the discovery, it became clear that x-rays were lethal. However, as is evident from the death rate of the early radiologists (Court Brown and Doll, 1958), the medical profession was not similarly constrained.

The medical profession owed its autonomy to claims of a scientifically based capacity to diagnose and cure, but by the end of the century there were few *in vivo* diagnostic devices and none as spectacular as x-rays. Radiographs also possessed the advantage that a group of physicians could discuss an accurate visual representation of a patient's anatomy without the patient even being present: this degree of objectification of the living patient meant that medicine could both be and appear to be more scientific, an important ideological benefit. Indeed, popular belief in the scientific effectiveness of the machines was so well established by the 1930s that patients took to demanding 'an x-ray all over' (Reiser, 1978: 163).

Medicine did not, however, have a monopoly over the medical uses of x-rays. The machines were developed by physicists and electrical engineers who, in the early years, commonly interpreted their radiographs for the medical profession. Indeed they were foundation members of the British Rontgen Society, later the British Institute of Radiology. Larkin (1983) argues that the participation of laypersons in the diagnostic process central to the practice of scientific medicine gave them a unique position from which to observe those diagnostic errors (and therapeutic failures) which could discredit it. The medical profession rapidly acted to exclude these fellow scientists from the medical applications of x-rays by the establishment in hospitals of their own physician-controlled x-ray facilities. In Victoria, the physicists who had introduced the technique to Australia had been replaced by hospital-controlled departments as early as 1898 (Eddy, 1946).

The end of the First World War, during which x-ray diagnosis was used in field hospitals, brought a flood of informally trained operators and simpler machines. There was a perception that the market could be swamped. In addition, by the 1920s, the deaths of the early pioneers from radiation-induced cancer (100 deaths by 1922) were undermining the claim of the medical profession to being able to control radiation hazard. The early 1920s therefore saw a number of important changes. Firstly, relatively effective radiation protection measures were instituted. These had the effect of substantially reducing the death rate among radiologists (Court Brown and Doll, 1958). Secondly, the medical profession instituted post-graduate training in radiology; in Britain their professional organisation was recognised in 1921. Later writers have claimed that this move occurred when it became clear that 'something would have to be done to train radiologists to serve the expanding specialty' (Shanks, 1950: 48) but we argue that the establishment of the specialty was a necessary part of the formal procedures used by the medical profession to licence itself as technically expert in the use of diagnostic x-rays. This in turn was a necessary step to appropriating the production process and establishing medical authority over the large number of other workers skilled in x-ray production.

The subordination of non-medical practitioners had two purposes: to deskill the lay workers by removing their diagnostic skills and to enhance the radiologists' own lowly status within medicine by shedding the production role (Larkin, 1983). This hierarchicalisation was justified by reference to scientific rationalisation; scientific efficiency called for the delegation of production tasks to lesser skilled workers while the medical profession retained those higher skills involved in the interpretation of the data (Reiser, 1978: 117). Such 'helpers' (radiographers) would be invaluable in helping the medical profession to satisfy the increased demand for x-rays but they would be 'extremely dangerous when they seek to practise independently' (Hernaman-Johnson, 1919: 81, quoted in Larkin, 1983).

Hence the second aspect of the professional's claim to expertise. The professionals, through training and experience, claimed a wider, less determinate understanding of a process which enabled them not only to supervise the execution of the technical aspects of the procedure by other, less trained workers, but also to interpret the results and integrate them into the clinical process. Of course, the claim to both esoteric technical skills and to indeterminacy is not a strategy adopted only by the medical profession but, where the medical profession succeeded in establishing its claims, other occupational groups such as skilled metalworkers have not. Indeed, when early radiologists laid claim to special expertise which they did not in fact possess, what was there to prevent other skilled groups like physicists and engineers from using the same weapon?

Larkin (1983) recounts how non-medical workers fought back by forming their own representative bodies to promote their own claims. However, when the British Society of Radiographers sought recognition by the Sate, its application was passed to the General Medical Council for comment. This body determined that radiographers should be recognised conditional on subordination to medical control. Radiographers, lacking the social power of the medical profession, were forced to submit or be excluded from practice. In this way the State effectively recognised the expertise of one group at the expense of the others.

What was claimed on ideological grounds was now to be achieved in fact; medical domination over radiographers' education boards determined that henceforth their training would be confined to the production role of execution, thus excluding them from the interpretive process. At this point, argues Larkin (1983), men with an autonomous occupational base like the electrical engineers, left what they acrimoniously called 'The Society of Radiologists' Assistants' and radiography became a largely female occupation. The work of radiography was seen as being peculiarly suitable to women as it called for them to have a 'real, deep, loyalty' to the radiologist. Only in 1960 did British radiographers gain professional recognition. Physicists and engineers, excluded from the diagnostic process, extended their own independent study of the applications of x-ray technology.

Against this background it is difficult to sustain the claim that technological rationality necessitated the deskilling and subordination of the job which became that of the radiographer. As one of us has argued elsewhere (Willis, 1983: 34), even if the machine created the need for technical helpers, this does not explain why there was a disproportionate number of women, working under the control of men at much lower rates of pay. Instead we believe that arguments of technological rationality and claims to professional expertise serve as the ideological justification for a licensing system based on State patronage.

The dominance of the medical profession over the practice of radiology received a setback after the Second World War when the aftermath of the atomic bombs used on Japan focused renewed interest on the control of what was now demonstrated to be an appalling hazard. Internationally, radiation control measures are now determined by the International Commission on Radiological Protection, a body heavily dependent on the expertise of physicists, especially that elite group involved in the nuclear industry (Radford, 1981). This body has raised serious questions about the capacity of radiologists to control the dangers to their patients from the use of unnecessary doses of x-rays and recommended in 1983 that certain common radiological procedures were 'dispensable' (ICRP 1982). Today, in hospitals, medical physicists have re-established themselves in radiology departments in a long overdue recognition of their skills in controlling radiation hazard.

Using arguments based on radiation hazard, radiographers in Victoria in 1979 also pressed their professional claims arguing, like the radiologists of 60 years ago, that 'ionising radiation is a highly dangerous tool in untrained hands'. The medical profession, they argued, not only took defective plates, but had a profit motive which resulted in the taking of unnecessary x-rays. For example, in 1979, 21 per cent of general practitioners x-rayed the pregnant abdomen. Public safety arguments, claimed the radiographers, were used by the medical profession to justify refusing radiographers registration when the real reason for the failure of their claims was that the Australian Medical Association could 'counter... responsible submissions by reaching the upper levels of government behind closed doors' (Hospital Radiation Technologists' Association of Victoria, 1979).

There is evidence from the United States that radiographers do take safer x-rays than medical practitioners (Wochos et al., 1979). However, the same study shows that untrained operators (orderlies and receptionists) also take safer x-rays than medical practitioners, but not as safe as those taken by radiographers. Despite the question this raises over the role of training, it is primarily these informally trained operators who will be excluded from x-ray practice. The pattern, it appears, persists; the labour process has developed around a pattern of deskilling less powerful users of a technology.

In 1983, with the *Health (Radiation Safety) Act*, Victorian radiographers finally won their battle for registration. It represents a substantial change in the attitude of the State that, in 1983, radiographers' claims could be granted directly by the State in the face of oppositional pressure from professional medical organisations. The change, as we argue in the next section, originated in the rising cost of health services.

Thus we argue that the x-ray machine has not itself created a new division of labour in health care, but it has facilitated the process by which those already in authority have reinforced pre-existing divisions based on class and gender. This authority of the medical profession over other occupations in the health field is legitimated by claims to special (indeterminate) expertise, but it originates in State patronage.

Medicine, technology and the State: the case in Victoria

By the 1950s in Britain there was growing concern over large rises in the number of x-ray diagnoses (Brailsford, 1952: 679). Not only were x-rays being used routinely (on one out of two hospital admissions: Shanks, 1950: 44), but there was growing doubt about the benefit to the patient (Reiser, 1978: 160). Nuclear physicists started claiming that unnecessary exposure to diagnostic x-rays in the United States was causing between 3500 and 29 000 cancer deaths each year (McClenahen, 1970: 453). The financial costs too were rising; by the 1980s the use of diagnostic x-rays in the United States consumed $7.6 billion each year of the $85 billion spent on hospitals (Richardson, 1987).

Radiologists in the 1950s and 1960s were concerned that, if they did not themselves regulate the use of x-rays, the State would force it on them (Brailsford, 1952: 679). In addition, patient demand for technological diagnosis was undermining the indeterminacy of their clinical skills. This latter problem was to be overcome by the development of spectacular new non-invasive imaging technologies which were to exacerbate still further the first problem: cost. The 1970s saw the advent of the CT scanner, a computer amplified x-ray scanning device which could produce vivid representations of cross-sections of the human body including the soft tissue, making it of particular importance in the diagnosis of tumours. But CT scanners are now dwarfed in diagnostic capacity, complexity of the image produced and cost by the latest in imaging devices, nuclear magnetic resonance (NMR) imaging.

The cost of CT scanners is high (about A$1 million) but at the time of writing there were 30 such scanners in Victoria alone, with a further twelve likely to be installed over the next year (*Australian Doctor*, 28 November 1986). Only eleven of the machines are in public hospitals. There is one NMR unit in a public hospital at an installation cost of about A$3 million. Since the capital costs of these machines can only be justified by maintaining a high 'throughput' of patients, the number of CT scans in the last five years has risen by over 400 per cent.

If we bear in mind that these new technologies only add on to the old technologies, rather than displacing them, then together they pose a clear threat to the health budgets of governments attempting to restrain spending. In such a context the radiographers' claims to being able to contain demand may well have won them registration even though it was granted under the ideologically more acceptable claim to promoting radiation safety. The same radiation safety legislation was used to cover the first 'certificate of need' legislation which allows the Victorian Department of Health to regulate the introduction of any radiation device over the value of A$80 000.

The reason why the Victorian government uses such indirect means to control the radiologists' 'tools of trade' is that State intervention in the medical field is strongly resisted. When the federal government decided in 1985 to regulate the use of private radiological services in its public hospitals by limiting the fee payable for such services, New South Wales doctors resisted with the well-known industrial tactic of going on strike. The dispute was resolved by the Commonwealth government agreeing to buy new radiological equipment for public hospitals and by setting up a commission of inquiry into the question of private practice. The Penington Inquiry concluded that the arrangements for private practice in public hospitals should remain because, in some cases, it was more economical.

Any attempt to place controls on the use of radiological 'high-tech' is met with the Royal College of Radiologists' argument that its concern is to 'maintain the highest standards of health care in the area of radiological sciences ... The issue at stake is community access to modern, proven diagnostic services'. Or, more emotively: '1000 people with curable cancer die each year because of a lack of proper equipment' (*The Sun,* 14 December 1984). Under such ideological cover, radiologists can apparently expand their services at will.

The conclusions reached by the Penington Inquiry show that government intervention is practicable when it is framed in terms of economic rationality. As argued by Duckett (1984), the State has established itself as a force in the Australian health sector using arguments based on economic rationality, that is, the need to constrain health care costs. Corporate rationalists located in the State institutions which fund health care services have developed technocratic solutions to health costs which are a direct challenge to the hegemony of the medical profession.

It was against this background that the Victorian parliament recently initiated a study into the labour market for radiologists. The government's stated concern is to correct understaffing in public hospitals with the drift of radiologists into lucrative private practice. A major finding of the inquiry (Economic and Budget Review Committee, Fourteenth Report, 1986) was that radiologists in private practice earn an average gross income of A$400 000, 65 per cent higher than that of other specialists. Such incomes were seen as coming from a labour market shortage of radiologists created by the labour-intensive new technologies, by a growth in demand, particularly in the private sector, and by a restriction of supply by the high failure rates in the radiologists' qualifying examinations.

The recommendations set out by the inquiry can be seen as a form of contestation over this labour process. Of most relevance was the recommendation that the appropriate authorities consider allowing radiographers to perform a wider range of services including the interpretation of the more routine diagnoses. Another recommendation was that the number of radiologists be increased by state government monitoring of the radiologists' entrance examination failure rates and by recruiting overseas radiologists. Thirdly, it was recommended private practice incomes be restricted by reviewing the fees payable for radiological procedures.

The radiologists have responded in classical professional style by describing the report as a 'series of bizarre recommendations ... some of which if implemented would seriously affect the quality of health care Australia-wide' (letter to the *National Times on Sunday*, 6 November 1986). The inquiry had already been advised that radiographers do not have the training to provide independent radiological services and would at any rate be reluctant to do so. The Royal Australian College of Radiologists (RACR) has also signalled that any such proposal would be strongly resisted. Its argument is simple but effective: 'You wouldn't want someone without medical training reading the results of your x-ray, would you?' The proposal for cutting fees it sees as unworkable (*Australian Doctor*, 28 November 1986). However, the economic arguments also have to be countered. The RACR claims that the government is trying to create a mood of public acceptance of government action against radiologists. It was unjustly blaming radiologists for 'excessively high' fees when in fact, they argued, radiologists spent sometimes over 90 per cent of their gross incomes on the operating costs of their practices.

In May 1987, the Commonwealth government struck the latest blow in this battle by reducing the radiological fees by between 5 and 25 per cent, citing as a reason the 'reduced cost of delivering services due to the introduction of high technology'. But it was described as an interim arrangement pending a review of services including the cost of producing them (RACR, 19 May 1987).

Conclusion

In this paper we have examined the applicability of concerns of the labour process tradition to indirect wage labour relations associated with 'professional' work. We have argued that the accounts of both ideology and the State must be more centrally incorporated in order to adequately do this. In the professional labour process, resistance is aimed at preservation of the indeterminacy of professional practice and of autonomy. The case study of the labour process surrounding imaging technology demonstrates how effectively this form of resistance through ideology can operate. There is, however, nothing intrinsically labour intensive about technological innovation in the health arena which would prevent the deskilling of workers including the medical profession. The preservation of autonomy and protection of a medically dominated labour process ultimately rests not on technology but on continued State patronage. In the face of increasing technicality of medical tasks associated with technological innovation, the emphasis on indeterminacy can be seen as an ideological rearguard offensive. In such cases the ability of professions to continue resisting deskilling depends upon the continuing articulation of their interests with those of the State. The other consequence, though, is the possibility of empowerment of other medical workers as the ideological nature of the claims to indeterminacy in the face of increasing technicality becomes more apparent.

9 Management strategies in the motor vehicle industry

ROGER WILKINSON

The motor industries of many advanced capitalist economies are currently undergoing major structural changes. Caused largely by the increased penetration of Japanese-produced vehicles into home markets, but also due to shrinking and volatile product markets, these changes mark a significant departure from Fordist principles of mass production:

> Instead of producing a standard car by means of highly specialized resources—workers with narrowly defined jobs and dedicated machines—the tendency is to produce specialized goods by means of general-purpose resources—broadly skilled workers using capital equipment that can make various models. (Katz and Sabel, 1985: 298; cf. Tolliday and Zeitlin, 1986)

Australia's role in this increasingly global industry is changing from that of a predominantly autonomous designer and producer of vehicles for the home market to an 'assembler' of components. It has long been recognised that Australia has an overabundance of manufacturers and suppliers of components. This fact, combined with a poor industrial relations record, has contributed to the low levels of productivity in the industry. Current government legislation aims to reduce the number of domestic vehicle manufacturers, urges closer co-operation between existing manufacturers and encourages the use of advanced technology and new manufacturing practices in the labour process to increase productivity. The Hawke Labor government's rationalisation plans for the industry are designed to subject it more and more to 'market forces' by gradually decreasing levels of government assistance. It is hoped that, under this pressure, the number of manufacturers and suppliers will be reduced. The industry has responded to these measures and is currently engaged in an intense search for a more efficient manner in which to conduct their operations.

This conjuncture faced by the industry provides an interesting setting to examine a number of changes in the labour process. The contemporary reorganisation of the labour process in the motor industry highlights a considerable degree of experimentation, and many of these experiments and programs challenge some of the basic assumptions underlying conventional management strategies for the control of labour. This study is concerned with two major areas of labour process theory. Firstly, many attempts to reorganise the labour process suggest that 'scientific management' and 'Fordism' may not be the most appropriate and efficient means of labour control. The amazing success of Japan in recent decades has forced Western management to rethink its attitude to workers and their contribution in the labour process. Secondly, the 'deskilling' thesis, popularised initially by Braverman (1974), has, in the light of recent research, been shown to be too simplistic. The introduction of robots, 'just-in-time' production, quality circles, multi-skilling and other attempts to reorganise the labour process all represent possible management strategies for corporations in this harsh struggle for survival.

This chapter examines this latter aspect, concentrating on changing production and management practices. It also considers the changing role of unions and the restructuring of jobs and skills.

The study

The material for this chapter has been gathered as part of an ongoing and larger study of the reorganisation of the Australian motor industry. The research began in 1983 and consists of a study of two US and two Japanese vehicle manufacturers based in Australia. A number of component suppliers were also studied. Interviews were conducted across a wide range of occupations, including corporate managers, middle managers, supervisors, workers and trade union officials. In relation to workers, tradesmen and unskilled workers, men, women and migrants were interviewed. About 250 people were interviewed and the interviews were semi-structured.

From 'just in case' to 'just in time'

Vehicle assembly plants are a complex and complicated array of manufacturing processes whereby thousands of discrete components are transformed into a finished vehicle. Components for this process can be manufactured by the company in-house, at another company plant or by an outside contractor. Inconsistent quality and wrong, incomplete and late delivery of components are quite common and, together with the possibility of the breakdown of key sections of the production process, can cause delays. Manufacturers as a result have, in the past, had to insulate themselves from these many and varied fluctuations in order to maintain consistent production schedules. Often through bitter experience, companies have developed the practice of having weeks of spare stock to off-set these potential fluctuations 'just in case things go wrong'.

The practice of having large stockpiles of components 'just in case' they may be required is expensive and contributes significantly to cost and productivity disadvantages between Western vehicle manufacturers and the Japanese (Sethi et al., 1984: 47). Slowly, following Japanese manufacturing practices, Western vehicle manufacturers are changing from 'just in case' to 'just in time' production. This new method of manufacture has important implications in two areas: the relationship between the vehicle manufacturer and the components supplier on the one hand, and the organisation of the assembly line and the labour process on the other.

Manufacturing vehicles in Japan may be just as chaotic, but much of this chaos happens away from the assembly line or outside the plant. Japanese vehicle manufacturers source a higher percentage of components from outside suppliers, many of whom are quite small in size and almost totally dependent on a single manufacturer for their livelihood. The manufacturers have been able to enforce a system of *kanban* (translated as 'just in time' production) whereby stock inventories have been reduced to days and, frequently, hours of components. The components are delivered by the supplier company to their required place on the assembly line *when* they are required. Otherwise the supplier loses their contract (Sethi et al., 1984: 47). Also, Japanese vehicle manufacturers have what many Westerners see as 'unhealthy', relatively open access to suppliers' book-keeping accounts to determine what they see as an 'acceptable' rate of profit. In the West, it is doubtful whether this state of dependency could exist, because many components suppliers are themselves multi-national corporations operating in oligopolistic markets. Despite this, many Western vehicle manufacturers have had qualified success in the implementation of 'just in time' production.

The 'just in time' system of production has also led to a drastic reorganisation of the layout of assembly lines and the placement and use of machinery. Where simple components are manufactured in-house, or when a number of components are formed into subassemblies, there is a growing tendency to locate these processes on the assembly line at the point where they are required and to produce just enough to enable the assembly line to operate. The manufacture of doors can be used to illustrate the different approaches.

The Ford Broadmeadows Plant in Melbourne plans to introduce the 'just in time' production system throughout its manufacturing operations, but at present door panels are stamped from sheet metal and sub-assembled at Ford's Geelong Stamping Plant some 100 kilometres away and transported daily to Broadmeadows. They are then off-loaded and taken by forklift and stacked in bins near their required position on the assembly line. Here they may sit for weeks before they are required. From the point of view of the 'just in time' philosophy, this practice is time consuming and costly, and little indication of quality standards can be gained until the components are fitted to the vehicle.

At the Mitsubishi Tonsley Park Plant in Adelaide the process of manufacturing doors is rather different. The metal-stamping presses required to stamp out the door skin and frame are situated in very close proximity to the operator who fastens and aligns the doors to the vehicle. At any particular time reserve units are limited to three or four. Apart from the benefits of less transport and handling costs, this method of manufacture makes it easier to monitor and maintain consistent quality products. Despite popular mythology, machines do not produce consistent quality products endlessly but need constant monitoring of tolerances and subsequent adjustment. While classified as 'unskilled', assembly line workers become very proficient at detecting minute changes in tolerances and quality. Hence those who fasten and align the doors can quickly alert those producing the doors if the quality diminishes.

While there are considerable gains to be achieved from the adoption of 'just in time' principles it can place considerable strains on the production process and also on workers. If management is serious about producing high quality vehicles and, following the Japanese example, accept the 'do it right the first time' attitude it can lead to considerable stoppages of the assembly line as production workers are given greater power to stop the line if it becomes increasingly difficult to maintain quality standards. This is related to the decision by many companies to reintegrate the inspection and the production processes, forcing responsibility for quality standards back on production workers.

While 'just in time' production has not been established long enough in Australia for a thorough assessment to be made, one significant feature has emerged. As the number of components manufacturers diminishes in line with the Hawke government's rationalisation plans, vehicle manufacturers are now finding that their production schedules can be seriously disrupted by a strike at a supplier's factory. Previously, vehicle manufacturers would discourage the formation of complete monopolies among suppliers by granting contracts for the supply of one component to two or more suppliers. Recently, a number of strikes at components suppliers, sometimes employing less than 100 workers, led to the cessation of production of a number of vehicle manufacturers within a matter of a few days.

In addition, as critics of the operation of 'just in time' production as it operates in the Japanese motor industry point out, it is not only stock inventories that are reduced to a minimum; staffing levels and work tasks are continually being reduced (Dohse et al., 1985: 131–2). In Australia it is difficult to make comparisons with Japan regarding staffing levels with the transition to just in time production. Regardless of 'just in time' production, almost all workers interviewed felt a greater burden of responsibility as they were increasingly required to perform both production and inspection tasks; frequently they were being asked to inspect the quality of their fellow workers' work before performing any tasks themselves.

Technological change

Assembly line work in the vehicle industry is often viewed as *the* most boring, monotonous and alienating form of work in modern times: '. . . the classical symbol of the subjection of man to the machine in our industrial age' (Walker and Guest, 1952: 9). It has remained relatively unchanged since its development by Henry Ford early this century through to very recent times. The typical image of assembly line work in the vehicle industry is that of a worker performing the same simple operation in a very short cycle time continuously for each minute of the work day, month after month. For consumers it has meant relatively inexpensive cars but, increasingly from the early 1970s, the quality and reliability of these products has been the subject of much concern and debate.

In the early 1980s the inefficiency of the Australian vehicle industry became of major public concern as corporate profits declined rapidly and many thousands of workers were subsequently retrenched. At the height of the debate about the Hawke Labor government's plan to restructure the vehicle industry, the *Australian Financial Review* (2 August 1983) noted:

> It is broadly accepted that the Australian motor industry is an inefficient manufacturer which is propped up by tariffs and import

> quotas; a manufacturer which could not survive if it was subjected to a harsh blast of competition.
>
> Up to a point, measures of efficiency such as cars produced per employee per year tend to support this view. Our "best" Australian motor manufacturer is perhaps Ford, which in 1982 achieved the rate of 12 vehicles per employee per year. Yet this was miserably less than the figure achieved by Ford's Highland Park plant for Model Ts in 1913 (19.3 vehicles per employee per year, and nowhere in sight of Japan's national average of 11 motor manufacturers in 1982 (39.1 vehicles per employee per year).

One of the inherent problems with the concept of productivity in terms of 'vehicles produced per employee per year' is that it tends to lay the blame for low productivity on lazy and truculent workers and promotes 'labour saving' technology.

The debate about 'new technology' is by now well known for its diversity. On the one hand it is claimed that the introduction of new technology will not only increase productivity and quality but will lead to better and more meaningful jobs, upgrading the skills of workers. On the other, critics see the introduction of new technology as a means to further control and deskill the workforce (Dunkerley, 1980). Unfortunately, much of the debate about the effects of new technology has been conducted at a high level of abstraction and highlights a lack of understanding about the actual workings of new technology in empirical situations; it is necessary to go beyond manufacturers' claims and marketing brochures. For example, the introduction of robots into the vehicle industry in Australia, as elsewhere, reveals a more complicated and less optimistic view than conventional wisdom would have us believe.

There is a common view that robots are in widespread use throughout the vehicle industry, that their rate of installation is rapid and increasing, and that new applications for robots in the production process are continually being discovered and developed. Robots can be seen operating in the spot welding, spray painting and assembly areas of most modern vehicle manufacturing plants. Robots, we are frequently told, are tireless, don't need lunch breaks and, most importantly, they don't *strike*. It is also claimed that robots produce exceptionally high quality products with unending consistency. Given these claims, it appears rather puzzling that such little use is made of robots in the vehicle industry, both in Australia and in other advanced capitalist societies.

One of the great attractions of robots for manufacturers is their flexibility and versatility. Conventional assembly lines typically produced only *one* type of vehicle and derivatives such as station wagons or two-door models. Retooling for a model change was an extremely expensive operation and entailed considerable down-time on the assembly line. On assembly lines that make extensive use of robots, a number of different models, or body types, can be assembled on a single line. In these cases, robots can easily be programmed to perform the required tasks by an operator who moves the robot arm through the sequence of tasks. This information is then stored in a computer which is linked to the robot. However, despite this flexibility, the use of robots is limited. Firstly, robots do not at present have the ability to be able to respond to unforeseen circumstances. Secondly, the use of robots is confined to handling, assembling, welding and spray painting. The 'power train' of the vehicle, a significant aspect of a vehicle including engine and transmission, requires substantial machining which is now performed in the most modern plants by numerical control (NC) machinery (Marsden et al., 1985: 39–40, 46).

One area of production that is increasingly being robotised in Australian vehicle industry is the spray painting of car bodies. The preparation and process of spraying car bodies is the most time-consuming aspect of vehicle production. Because of the need for high quality it also causes major bottlenecks in the production process. Companies measure quality control in this area by the 'rework rate' (i.e. the percentage of cars needing additional attention to reach a required standard). With experienced manual spray painters, most companies estimate that a 30 per cent rework rate in this area is acceptable. Furthermore, experienced spray painters are costly to train and difficult to retain. With all the claims of unending performance, increased productivity *and* quality associated with robots, it would seem logical and appropriate for management to introduce them into this most critical section of the production process. However, the introduction, operation and performance of robots has not as yet provided the improved productivity and quality that was expected. For example, one company that had made significant improvements in productivity and quality in vehicle production, and had implemented robots in the spray painting over two years previously, claimed that it was still experiencing 'teething problems'; the rework rate was frequently 50 per cent and occasionally 60 per cent. Another vehicle manufacturer claimed a slightly better record for its robotised spray painting facilities. The spray painting of vehicles is a highly critical operation in the product process and quite a complicated one; given the problems experienced, the reluctance of management to use robots becomes more understandable.

Of course, it is easy to claim that these results 'are only to be expected' in this complex area of the production process, but even when performing more simple operations the many expected benefits of robots are not always forthcoming. This can be illustrated by considering a study of the printing industry:

> the company put in a robot that was supposed to grab loose newsprint sheets as they came off a printing press. When the robot arrived, the employees were curious. They expressed concern over potential job loss and openly predicted that the robot 'would never be able to do this work as well as a man'.
>
> They were right. For more than six months the company tried to debug the robot on the factory floor. In all that time no employee suggested any way to make it work. Finally, . . . the company removed the robot. Had the company chosen a simpler task for its first robot, workers might have responded with enthusiasm. (Foulkes and Hirsch, 1984: 96)

In another example, an Australian vehicle components manufacturer management introduced a robot to remove sand from hot castings. In this operation a worker would give the casting 'a couple of taps' until the sand was removed. The robot was taken through the human motions entailed in this process but, once in operation, proved unsuccessful because it often required 'more than two taps to remove the sand'. The robot was later removed. These two isolated, but by no means uncommon examples, highlight some interesting points. Firstly, in the two examples both jobs were considered by management as relatively simple jobs that could be performed by 'unskilled' labour. Yet in the first example the authors argued that management should have chosen a 'simpler task' to robotise! This would seem to suggest that robots and advanced automated machinery, such as NC machinery, are not as sophisticated and reliable as their marketing and advertising implies (cf. Melman, 1983: 106–9). Also, even 'unskilled' workers develop skills on the job that are indispensable and vitally important for the effective operation of the production process (Juravich, 1985; Jones and Wood, 1985). Secondly, in relation to the example from the printing industry, the authors also complained that workers were not forthcoming in suggesting possible solutions to the problem. This is not to suggest that workers would not be able to offer possible solutions, but that bitter experience would have taught them that managers would not take their suggestions seriously.

While the 'technical' problems associated with the introduction of robots go some way to explaining the reluctance of management to employ robots in the vehicle and other industries, the costs associated with employing 'human' labour should also be considered. Australia is often considered a 'high wage, low productivity, strike-prone country' and therefore robots are often seen as an attractive and alternative proposition to employing workers. However, as Prime Minister Hawke recently told a delegation of visiting Japanese businessmen, the image of Australia as a 'high wage, low productivity, strike-prone country' is a thing of the past (*Australian Financial Review*, 10 February 1987). The award wage for unskilled production workers in the vehicle industry in late 1987 was $244.00 per week, hardly a high wage. There are, of course, other compelling reasons for employing human labour in the current period of uncertainty in the Australian vehicle industry. 'Direct labour', that is, the costs of employing production workers, accounts for only a small proportion of total costs (*Business Week*, 13 February 1984; Skinner, 1986: 56). Also, the number of workers replaced by robots is frequently overstated and is usually calculated on a three-shift basis; when calculated on a single-shift basis, which is more normal for most Australian plants, much of the labour replacement value is diminished (cf. Hartley, 1983: 36). Therefore, it would seem to make little sense to allocate large financial resources on labour replacement technology in a period of dramatic restructuring and uncertainty. Workers would appear to be a more attractive proposition because they can be hired and fired much more easily.

Changing management practices

While the production process has changed dramatically in recent times (even if the extent and effectiveness of advanced automation may have been overstated), a good deal of time, money and effort has been spent on changing long-established management practices. Many management positions have been reduced, especially lower management tasks such as supervision and inspection, not only as a result of massive reductions in the number of production workers employed but also because of the restructuring of the production process. Also, following the success of Japanese vehicle manufacturers in combining productivity and quality, Western management has begun to introduce Japanese management practices such as quality circles. By the mid-1980s, all the vehicle manufacturers and many parts suppliers had tried quality circles. These should not be confused with a related innovation, 'quality of work life' programs, which are largely confined to Ford ('Employee Involvement') and General Motors ('Quality of Work Life').

Quality circles (see also Chapter 6) are introduced with a variety of aims such as improving product quality, improving labour productivity and increasing employee participation and commitment to the firm. However, production-related problems usually take priority over motivational aims. Quality circles are not encouraged to tackle other award matters—problems conventionally addressed by trade unions, or management concerns such as investment and long-term planning. Quality circles are founded on the assumption that 'the person doing the job knows far better than anyone else the best way of doing that job and therefore is the one person best fitted to improve it' (*Fortune* editors, 1982: viii). Quality circles aim to identify production problems, suggest likely causes, gather data on the extent of the problem, highlight alternatives and suggest a solution which culminates in a 'management presentation'.

Each quality circle group in the vehicle industry consists of somewhere between six and twelve workers, who work in close proximity and meet on a regular basis, usually weekly, to solve production problems related to their immediate work area. The smallness of the group is designed to provide an atmosphere that will make even the most restrained person feel comfortable to voice their opinion. The idea that group members work in close proximity is important not just because they know one another and usually face similar problems, but in the longer term it avoids conflict throughout the factory and enhances the life of quality circles. For example, in the early 1980s many early attempts at quality circles operated on the assumption that wherever you had a small group of workers sitting around a table, then this constituted a quality circle. In one factory a quality circle was formed drawing workers, supervisors and foremen from different sections. One of their projects was to investigate why assembly line workers were frequently running out of parts. Discussions continued for three months, but the problem was never resolved because each section privately blamed other sections. According to one of the members, 'no-one would go public'. Comments included: 'I couldn't have a go.' 'I would have liked to have spilled the beans.' 'I pulled out.' 'I told the facilitator I couldn't speak the truth.' This pilot quality circle had a very limited lifespan and caused a good deal of friction and disappointment. Its demise signalled the death of quality circles in this particular factory. This is why it is considered important that the members of each group work close to each other.

Another important element of quality circles relates to the status and power of individuals. In Japan, workers bring into the group their power and status from the shopfloor. Furthermore, people in positions of authority, such as forepersons, supervisors and leading hands, become permanent leaders and actively participate in the direction of the group and encourage competition between groups. In Australia, the more serious and long lasting quality circles aim to be ultra-democratic. Group leaders are elected for a specified time and anyone may become a leader. Also, members come into the group as equals, so that, in theory, an assembly line worker has as much say as a supervisor. In practice, superiors tend to have 'the gift of the gab', as one quietly spoken member put it, and sometimes dominate discussions. Many quality circle members believed that they would achieve more without their superiors, because they could 'open up more'. Leaders of groups are encouraged to assume a passive role, co-ordinating and chairing meetings rather than actively dictating the direction or agenda.

The training that quality circle members receive helps them to set their own agenda and solve their own problems. Members spend many hours over a number of weeks learning problem-solving techniques. The aim of the training and the rigid application of the techniques is used to systematise problem identification and solution.

Problems arise through the process of 'brainstorming'. Initially, members are asked in rotation to identity *problems* or *symptoms* related to production, no matter how trivial or how large. At this stage there is no evaluation of either the cause of the problem or the worth of the problem. The aim here is to encourage workers to 'open up' that 'locked away' information which they possess. The first brainstorming session often highlights 40 to 50 problems and takes two or three one-hour sessions to complete. While the group is encouraged to tackle production problems related to their immediate work area, this is open to interpretation. Some suggestions are highly specific, such as 'catalogue part numbers and a stamp to be imprinted on to weld jigs for easy identification.' Others are more general, such as 'arc welders to be replaced with modern equipment' or 'modernise paint system in seat section'. In the normal course of events, quality circle members would be directed away from these two problems as they relate to investment decisions well beyond the scope of quality circles.

Quality circles are urged to start with small and manageable problems and work up to bigger and more complex problems. A popular starting point is always 'housekeeping'—the physical appearance of the factory— and an initial project might be 'to put parts in numerical order and establish a pegboard displaying parts.' Even a simple task such as this can take three or four months from the time the problem is identified until the solution is implemented. More complex problems take much longer. In one quality circle formed in the press shop at a vehicle manufacturer, the group was concerned about the growing incidence of dies falling off their tables because of vibrations prior to their installation in the presses. Dies are large blocks of shaped metal that are used to press body panels.

The circle reported that 'the personnel required to check controls on presses were at risk of injury.'

In one three-month period, four dies had fallen, but luckily no-one was injured. The solution was simple enough: bolt suitably strengthened brackets at the end of the 'load tables', thus preventing the dies from falling. Yet, while the solution seems simple, especially to those involved, the whole process from problem identification to installation, to management presentation, took well over *six months* as is illustrated in Table 9.1:

Table 9.1: Quality circles: life cycle of project

Selection problem	3 weeks
Collect data	6 weeks
Attempt countermeasure	3 weeks
Fit and reinforce countermeasure	3 weeks
Verify results	2 weeks
Install on [all] presses	5 weeks
Inform management	4 weeks
Total	29 weeks

Source: Company records

At public discussions about the merits of quality circles in Australia there is usually a good deal of debate as to whether they should 'pay their way'. In the example shown in Table 9.1, the facilitator argued that companies have to make a 'philosophical commitment to circles as a way of doing business. If you're going to make them pay then they won't get off the ground.' In this company's standard 'Quality Circle Project Report', there is no space where savings and costing can readily be entered. Still, other companies are insistent that they will 'pay their way' and urge quality circles to draw on accountants and engineers to consider 'the money angle', as one official put it, and are relatively satisfied with the results. However, as a general trend, companies in Australia, the United States and Britain do consider the cost of implementing quality circles and their likely returns (Bradley and Hill, 1983: 3).

In recent times, quality circles have lost some of their appeal. As a method of solving production problems, quality circles are extremely slow. From the point of view of quality circle members, they are seen as extremely ritualistic and stifling of enthusiasm, leading to frustration and 'quality circle burnout'. Furthermore, because of their slowness, quality circle groups sometimes only make one 'management presentation' per year. Some groups studied had not made a management presentation in over two years, despite the fact that they had successfully solved a number of problems. It is at the management presentation where, according to members, 'it all comes together', giving a 'great sense of satisfaction and achievement'. One of the reasons why presentations are so rare is that many managements have developed a practice of having a 'management presentation day'. In these instances, quality circles seem to have lost much of their appeal to management.

Quality circles are also extremely expensive. Currently it costs a company about $60 000 to establish *one* quality circle and maintain it for the first year. This cost includes the consultants' fee, the training of a facilitator and the training of the quality circle members and their lost time on the assembly line. Also, the pay-back period is estimated to be between three and five years. What is evident, then, is that quality circles in the Australian vehicle industry are very costly to implement, they are slow and cumbersome, and their success is not at all guaranteed. As a quick way of increasing productivity, quality circles are not really a viable proposition. Some have already prophesied their downfall (Lawler and Mohrman, 1985).

The reorganisation of jobs and skills

Since the late 1970s, the volatility of product markets and the formation of specialist markets have forced the major vehicle manufacturers to adopt a much more flexible manufacturing system than conventional 'Fordist' assembly lines. Some, if not most, of the major changes now being experienced are being felt by tradesmen. Tradesmen in the vehicle assembly plants are primarily concerned with the *maintenance* of equipment and machinery. Traditionally, there has been a fairly strict division of labour in maintenance functions and the repair of malfunctioning machinery may require several distinct trades. For a long time this was seen as a cumbersome and time-consuming method of operation, but inevitable, given apprenticeship training and industrial awards which specified in detail the demarcation lines of the relevant trades. Of course, the repair of machinery is not so easily susceptible to the principles of scientific management and, as Michel Crozier (1964: 108–10) has argued, this has allowed tradesmen to increase their political power because they have 'control of uncertainty'. Also, when tradesmen are working on a particular piece of machinery over a period of time, they develop a 'feeling' for what might be wrong and can recognise defective machinery by intuition.

A good deal of the new machinery now being introduced in the motor industry operates on pneumatics and hydraulics and many tradesmen don't necessarily have competence in these areas. Hence management has started to use outside companies to service and repair this machinery. In some cases this has led to a reduction in the number of tradesmen employed, leaving the existing tradesmen to the less sophisticated and more mundane aspects of preventative maintenance. As well as the contracting-out of many maintenance functions, there is a growing trend for manufacturers of new machinery to offer service contracts and supply maintenance tradesmen. In these cases, the tradesmen are employed by the machinery manufacturer but may work full-time in one factory.

Conclusion

It is evident that the Australian motor vehicle building industry is currently experimenting with a variety of management strategies to improve the competitiveness of the industry through increases in productivity and quality. The labour process, especially in this industry, is complex and whether these management strategies are successful is, in the light of the above discussion, by no means guaranteed. It is difficult to predict the nature of national markets for motor vehicles and the move towards the greater flexibility of the production process and of the workforce will certainly enable the Australian industry to adapt more quickly in this period of uncertainty.

The role of workers and their trade unions will also be important and a consideration here is the co-operation between trade unions in the industry, especially around the issues of jobs and skill. Tolliday and Zeitlin (1986: 21) suggest that skill requirements in the industry are rising and claim that how trade unions handle this issue is of vital importance. If they resist change they may be seen as contributing to the decline of the industry; if they accept change, then existing trade union structures will inevitably feel the pressure to change (Katz and Sabel, 1985). Whatever happens, the next decade is likely to see fundamental changes taking place.

10 The fast food industry

ERN REEDERS

In 1978, Australians spent a billion dollars on takeaway food. In the previous nine years, the number of takeaway establishments had increased approximately threefold and the number of workers in the industry increased approximately fivefold (Australian Bureau of Statistics, 1974; 1981).

An increasing proportion of families' budgets is devoted to the purchase of prepackaged or takeaway food, in keeping with the general trend towards more and more needs being met with commodities rather than self-made goods. Jones (1982: 79) reports that the *increase* in employment in eating and drinking places in the US from 1973 to 1980 was greater than *total* employment in the car and steel industries. In Australia, during the decade from 1971, the increase in jobs for males in the fast food short-order cook category amounted to 278 per cent, and there was also a large rise in demand for waitresses and barmaids (Windschuttle, 1986). In the future we can expect to see continuing increases in sales of takeaway food cooked on the premises, and also of food from vending machines whereby the principle of the human service industry (the immediacy of production and consumption) is replaced by factory production.

In this chapter, the focus is on the takeaway or fast food industry, characterised by fast service, a limited menu, standardised quality and low price. This includes fast turnover eat-in as well as takeaway establishments, and covers such edibles as hamburgers, fish and chips, pancakes, pizzas and chicken. The analysis will concentrate on the fast food chains.

A marked change in the industry has been the arrival of large multinational corporations. According to its promotional literature, McDonald's was established in Australia in 1971 and now has 183 outlets here and some 18000 full-time and casual employees. The company claims that it has over two million customers a week, and sells nearly 100 million hamburgers a year. Its sales in 1985 amounted to $232 million, an increase of 19 per cent over the previous year. McDonald's had a worldwide turnover of ten billion dollars in 1984, with 8304 restaurants in 34 countries. Kentucky Fried Chicken operated 6396 restaurants in 56 countries in 1985, with a claimed turnover of 2.8 billion dollars (Labour Research, 1986).

The industry is now split between the large chains and a myriad of corner shop operations. The corner shop usually comprises a working proprietor and family, including perhaps casually employed helpers. Working proprietors amount to nearly a third of all workers in the fast food industry (Australian Bureau of Statistics, 1981). The working family model is spreading among the chains as well, as they turn increasingly to franchise arrangements (in which owner/operators manage a store and pay license fees to the chain).

Some observers of production technology have argued that technological advances promise to relieve workers of the brute and boring tasks of labour which characterised industrial work in the past. Blauner (1964), for example, saw in the advanced continuous process technology of the oil industry a sign of the shape of work in the future: the new workers would be white-coated technicians monitoring gauges in sophisticated control rooms, with the status, qualifications and privileges of 'staff'. In a similarly grand vision, post-industrial theorists such as Daniel Bell (1973) have argued that a move to a service economy will usher in an age of more satisfying and skilled work as employees deal face-to-face with clients in helping to meet their needs and that, in the long term, white-collar work in the service sector would overtake the manufacturing industry in its share of employment. Implicit in the thesis of both authors is the view that many industrial jobs belonging to the secondary labour market (i.e. that part of the labour market where work is unskilled, insecure and offers few career prospects) would vanish in technological and economic change and be replaced by growth in the primary labour market where workers are skilled and qualified, have job security and opportunities for promotion (for a critique of Bell, see Kumar, 1977; and of Blauner, see Eldridge, 1971). It is in the context of visions such as these that we can examine the nature of work and the design of technology in the rapidly growing fast food industry. We must ask whether the fast food industry and the service sector of the economy in general *are* providing skilled work for full-time employees.

Available evidence in the case of the fast food industry indicates that, in this part of the service sector, most employment is in fact casual in nature (see Table 10.1).

Table 10.1: Employees in the fast food industry, 1978

	Cell percentage	
	Males	Females
Full-time	13	22
Part-time	17	48

Note: n = 33,592. ASIC class 4886, excluding working proprietors
Source: Australian Bureau of Statistics (1981)

In a typical McDonald's store, with from 60 to 90 people on the roster, only about 15 per cent of the staff are full-time. Part-time employment has become more common in other areas too—in retailing in general, and also in banking (Sweet, 1983; Macklin, 1982). It reflects not simply a way of dealing with daily cycles in demand, but also changes in the way work has been organised. What could have been full-time career-grade positions have been fragmented and parcelled out to low-paid casual workers. These workers are usually married women or juniors. In the UK, three-quarters of McDonald's workforce is under the age of 21 (Labour Research, 1986). The majority of teenage part-time workers are also school students and the proportion of teenage workers who are part-time has been increasing in leaps and bounds. In 1981, one in five male and one in four female teenage workers were part-time (Sweet, 1983).

As indicated in Chapter 1, since Braverman's (1974) seminal work on the labour process, a number of debates have taken place and shortcomings have been identified in labour process analysis. One of the issues has been whether deskilled or unskilled work is necessarily experienced as degrading work, particularly by workers who have never had the experience of skilled work (Thompson, 1983). Most of the employees in the fast food industry fall into this category and, indeed, following Braverman, belong to the category of new workers who have to be habituated to the capitalist labour process. Two shortcomings which have been noted in Braverman's work are his neglect of gender issues and of resistance by workers to managerial control.

The study

The accounts of some of the young workers in the fast food industry can be examined to throw light on some of these issues, and in particular to see whether the work they do: (firstly) requires and entails skills to meet customer needs in a flexible one-to-one service relationship: (secondly) is found to be intrinsically rewarding; and (thirdly) is structured along gender lines and offers scope for resistance.

With high rates of youth unemployment in Australia, the transition of students from school to work has been widely studied in the last decade, and one of the consequences of this has been an increased focus on the youth labour market and particularly on teenage part-time work. One of the issues in debates about youth unemployment has been the adequacy of school's preparation of students for work in terms of their abilities and their attitudes. The findings reported here form part of a larger study of the transition of students from school to work. In 1982, I interviewed at length 41 male students and conducted group discussions with twelve of them. Once they had entered the labour force three to four years later, I interviewed them again. The initial interview elicited accounts of their part-time work, of which nearly all had had some experience. Six had worked in the fast food industry. For this paper, I interviewed at length two further fast food ex-employees. 'Alan' had been with a chain for five years, ending up as a full-time assistant manager, while 'Joyce' worked as a casual for three years and became one of the chain's fastest register 'girls'. While most of the quotes here are from these two respondents, their comments are corroborated in the accounts of others in the study and are consistent with findings on the fast food industry elsewhere (see Rieter, 1987).

For the fast food multinationals, the opening up of a mass market by offering convenient low-cost food in a way which is necessarily labour intensive has meant stringent control over the way their employees labour. The reduction of labour costs has been a key problem for fast food operators. In the small family operation, it has been achieved by exploiting the family's own labour power (by not paying for it at market rates). Such an option is not available to the outlets of the large chains and so it is in some of these that all the techniques of control of human labour for ulterior ends have been applied. These chains have adopted a marketing strategy of projecting a clean, efficient and youthful image. Many thousands of teenagers, however, also get to see behind the image in their first experience of part-time work. Youth is at a premium:

> '... the cheaper they were, the more profit they were making. It madc a real big difference. If you had the younger staff on, you made a lot more money. Simple.' [Alan]

Most casual teenage workers are aged fifteen to seventeen (Sweet, 1983). Young women employed under the award for waitresses receive the same rate up to eighteen years old while the rates for young men working as cooks rise annually. Older workers mean less surplus value. Joyce said that, as employees got older, 'they never fired you, they just didn't give you shifts. So eventually you got jack of that, and left.'

As Table 10.1 shows, by far the largest single group of employees in the industry consists of part-time females. In addition to their skills in light repetitive work on the cash register, in one chain their appearance was part of the marketing strategy. Joyce recalled 'at our store they liked having girls at the front, and the managers got strict instructions that the girls had to be good looking. Or else they won't get hired.' Alan confirmed this: 'Yep . . . "We want attractive girls at the front counter. We wanna bring people here, not scare 'em away".' The appearance of the girls is thus harnessed for the purpose of attracting customers, though this strategy is now running into conflict with the task of cheapening labour as younger males are cheaper to employ than 'waitresses' and are starting to appear at the registers. Girls remain concentrated at the bottom of the chains' hierarchies. None of the young adults interviewed for this paper recalled seeing a female in a managerial position. McDonald's promotional literature, however, makes reference to the existence of one female manager in its chain. The respondents in this study also indicated that girls were more docile.

Low production costs and low labour costs especially are a prerequisite for mass marketing fast food as a commodity. But cheap labour alone does not guarantee that surplus value will be produced. This requires that workers produce value above that of their wage. For work done for others which is not worth doing in itself, or which is done by workers who are not fully subordinated to the production process (new workers, for example), a surplus can only be extracted through management's use of a coercive system of rewards and punishments in order to maintain consistent and intensive labour. The coercive system *internal* to the large fast food operations is bolstered *externally* by the high demand for casual work. Joyce recalled: 'that's what they used to say to you, to managers and to (staff): "For your position, there's five people wanting your job."' Alan confirmed this: 'It's true. There's so many kids wanting to get a job . . . They can *afford* to be sacked.' With this threat, we see the most basic form of coercion, coercion through the wage, achieved through the threat of being left off the roster:

> If they had an excess amount of (staff), if someone was sick
> they'd punish them by not giving them any shifts, for the next two

> or three weeks they didn't find themselves on a roster. But if they were busy and they needed the staff, of course they wouldn't do that. [Alan]
>
> If they didn't like you, personal, they wouldn't give you shifts. Yeah, it was all just threatening with shifts. [Joyce]

Rostering of staff is the key political process in the large chains. It is the tool for the exercise of wage coercion. But one of the contradictions of employing juniors, especially school students, is that most of them are not breadwinners, so what is threatened is often not their livelihood but only their pocket money. A second and related contradiction is that labour turnover among teenagers is high, making it difficult to fill the roster at busy times such as school holidays. Wage coercion is most powerful when applied to breadwinners, especially needy breadwinners. Alan observed:

> When I was leaving they were employing a lot of Vietnamese. Cos they were eager to make a buck. An' they were prepared to do anything. Long hours. They used to get abused. The verbal abuse was incredible.
>
> Lot of the Vietnamese kids, the boys anyway, they were supporting families in the high rise. [Joyce]

An important advantage in employing juniors is that they are generally ignorant of their rights. This even applies to some managers, many of whom, at the age of 21 or 22 years, are not much older than their staff and with only minimal training are also often ignorant of award rules and conditions:

> There's no union. I think it ought to be there . . . Simply because they're dealing with minors who don't know the law, and you can easily pull the wool over their eyes cos they don't know different. And that's what they used to do, you know, constantly . . .
> . . . When I was a manager, I was telling (staff), you know, to go ahead and have a (break). I didn't know what the story was, I just used to make things up as I went. That's what you'd find most managers were doing . . . Say it off the top of their head to suit themselves. (Staff) weren't gonna argue. [Alan]

Joyce said: 'You'd never know when you were supposed to have a food break . . . The ten minute break, they'd make it out as if it were a favour, but other people said it was a right.' She recalled occasions when a five- or six-hour shift was worked without a break or time for a meal.

Exact task specification is the rigid framework of the fast food labour process:

> You actually had a bench, with all your condiments, your pickles, whatever you had, all lined up. And you had a sequence in which you actually arranged everything. [Alan]
>
> Just like a conveyor belt. [Joyce]
>
> An' you'd get your buns out which'd been toasted by someone else. An' someone else's responsible for dressing the buns. [Alan]
>
> An' you'd spend eight hours, or your whole shift doing that. [Joyce]
>
> Oh it used to alternate sometimes. [Alan]

This quote indicates the degree of detailed division of labour evident in some of the large chains. In the separation of conception from execution, workers in this service industry perform repetitive unskilled tasks in a system modelled on the assembly line. Like the line, the pace of work is not decided by the worker, but in some cases by the clock, and in others by management. A routine is laid down and the inexorable flow of the product compels the worker's attention to the task. The food is assembled rather than crafted:

> Instead of putting the bit of fish on the bun first, I put the tartare sauce on the bun first, then put the fish on. (The manager) started to rave on about that and then he goes, 'How long have you been working here?' and all this stuff. I just said, 'Too long.' He goes, 'Okay, go and clock off and go home.' [David]

Separate assembly tasks can be learned in a few hours or days. In the chains, the form of managerial control is largely bureaucratic (Edwards, 1979). Everyone has a clearly delimited role to play in the hierarchical organisation. Joyce remarked that it was a conveyor belt. Alan said 'Firstly you learn the basic things. Then repeat 'em. Repetitive was all it was.' Joyce was even told how to mop the floor in the approved manner.

In one chain, an unvarying six-point routine for dealing with customers is demanded: greeting, taking the order, suggesting a further purchase, assembling the order, taking the money, and farewell. Company staff posing as customers check on the performance of register staff and the quality of food and service. Defaulting staff are reprimanded before their fellows and threatened with dismissal if performance does not improve.

There are also elements of technical control appearing in the industry, with the performance of tasks being governed by timers, conveyor-style pizza ovens and automatic microwave ovens:

> you couldn't do anything by yourself, it was all timed and stuff like that you know. Everything is timed to a second. You have got buzzers when you have got to take the hamburgers off, you

> have to press them when you put them on, and you have a buzzer that tells you when to turn 'em, when to sear 'em. [David]

By contrast, simple control typifies the corner shop operation (Edwards, 1979). Tasks are assigned more flexibly and there is much less fragmentation of work. An employee will often see a process right through taking an order, cooking, wrapping and receiving payment for it. A variety of skills are exercised, and there is scope to develop a personal style of work. Food can be cooked to suit the customer's preferences, a feature which is much rarer in the chains, since it disrupts the exact specification of tasks. Personal relationships can develop:

> first, second week I wasn't very sure of meself, but then I started going alright. When I started, you know, I was too scared to put the fish in the batter and then put it in the oil. But now it's just so easy for me, just go on, get fish, scallops, anything. . . Oh sometimes it gets boring when it's not busy and you have no work you know, and you just sit there doing nothing . . . There's four of us. The boss, the wife, me and now I've got my sister a job in there working on the till. So it's nearly a family affair . . . It will probably help me to get a better job. Like, I got counting experience with money . . . I'll get a reference from the boss. I've got experience with orders. Putting the right orders. Um, talking to people. Like everybody used to tell me when I just came back, the other bloke wouldn't really talk to anybody. But as soon as I came back in there there was more business. You know, cos I used to talk to everybody. But this bloke he just took the orders and sat down. I used to stay there and talk to everybody. [George]

In the chains, even the talk is standardised; register staff have to follow a prescribed routine in greeting customers and taking orders.

The intensity of labour is a key concept in labour process analysis. It refers in general to how hard and fast employees work, and in particular to whether and at what rate they produce surplus value. If George's experience is typical, then the intensity of labour in the corner shop operations is relatively low. This may be compensated for by family members working very long hours. In the chains, however, nearly all labour has to be paid for at standard rates. Hamburger assemblers may do exactly what their job specification requires but at a rate too slow to produce surplus value. Detailed division of labour by itself cannot guarantee a surplus. The task for management is to control labour in other ways as well, in order to achieve this. Alan observed: 'Its very easy to learn the basics, but what they expect you to know, I mean, is speed.' Joyce: 'That's the skill they're after, speed.' Alan: 'Anyone can put a hamburger on a hotplate.' Joyce: 'That was the criteria for whether you were a good (staff) or not, how fast you were.'

Maximising the intensity of labour is especially important at the low-cost end of the market where average transaction values amount to only several dollars. Maximum turnover with minimum staffing is required. Some of the pressure to work hard is mediated through the apparently neutral device of work organisation; at rush times, the pressure of a queue of customers is felt at the registers and relayed back to the cooking areas. Pressure also comes directly from the manager. Managers are expected to drive their staff:

> Specially in the . . . store, it was very busy, and you just, you know, worked your butt off. And managers treated (staff) like dirt. [Joyce]
>
> Yeah, I agree. We were really expected to treat 'em like that. But yeah, well, being young, most of the managers are fairly young, well, I mean, you sorta got this power craze. Really pushing your weight around. Really abusive. [Alan]

In one busy outlet, at rush time the 'girls' on the register went through their six-point routine as often as twice per minute for as long as four hours.

> you were programmed that you couldn't stand still. Even in really quiet times you had to clean. And, like, the bludge job was wiping down the trays . . . You always had to be moving. You weren't allowed to stand still . . . They hated you standing still. [Joyce]

To cut labour costs, managers were under pressure to keep staffing to a minimum or to deliberately under-staff. To cover the ensuing shortfall in effort and service, they might then rush around the shop floor pitching in themselves and offering bonuses for the register operator who smiled the most or who took the most money.

> It was all calculated. They'd put two girls who were fast, and like I was sucked in, put you with somebody else who you might not like, or there was rivalry or whatever or who was fast as well, they put you next to each other so that, you know, you got thousands of people in the store, and you've got the managers yelling all the time 'Come on girls, come on, we can break it', break the record for the hour. And the hour was up and they'd get the big roll of how much we'd taken and it was either hooray or no, and then we'd have to try again for the next hour . . . So you were conned . . . They were yelling it out. Don't care about the customers, 'Come on girls, ten dollars for who is smiling the best.' [Joyce]

Stores competed with each other for the highest turnover, and staff engaged in organised competitions between stores for the highest speed. Joyce joined in one of these competitions, and found that no wages were paid for the first heat.

While the same kinds of negative sanctions appear to be employed against both sexes, the positive sanctions applied in the intensification of labour are gender differentiated. While young women may get bonuses for smiling, many young males in the chains are induced to work harder by the prospect of a managerial career. It offers, apparently, a secure and respected position, the acquisition of new skills and a series of promotion steps. However, it may not be that much of a reward:

> You work your way up to manager but you're simply just another (staff) person, really, just someone who cops the blame. It's not like a manager, say, walking round with a case and your tie and your company car. Simply there just wrapping burgers, and you know, copping the flak. [Alan]

Manager's work, like the staff's, is all laid out: bookkeeping, reporting, rostering and, in one chain, responsibility for 'waste control'. This involves examining the garbage in the bins in the cooking area, estimating the quantity that has been wasted during the shift. Alongside this, the managers are given motivation tapes, go to conventions and meetings at (non-fast food) restaurants, and are urged by the offer of prizes to increase store turnover. They work long hours—often several weekends in a row or late-night shifts followed by morning set-ups less than eight hours later.

Like the staff, the assistant managers and store managers are themselves subject to coercive devices:

> I just didn't like the way I had to handle everyone . . . Lotta the staff were my age, and I was puttin' shit on 'em. And I just got really slack. Used to let people take smokos. Used to always make the money, used to run a pretty good shift I thought. But in a different way from the way they wanted me to. So I used to get reprimanded all the time . . . But a lot of managers think they're going to become a lot more than they ever were. Just really work hard, an' get nowhere. [Alan]

However, the *modus operandi* on the one hand and the objectives on the other often contradicted each other:

> They expected to be running a minimal amount of (staff) possible, making the most money they can, and serving the best product, which never eventuated. If you didn't have the staff, the product just wasn't right, service wasn't right. People just arguing. It was never right. [Alan].

Staff cuts and increased pressure were countered by attempts at 'morale building'. The contradictions were resolved in a time-honoured way:

> The quality control during the holiday periods . . . you just couldn't believe it. Supervisors were standing there just turned their heads so they didn't see it. But it's all for profit so they don't care. Had they come in on a quiet period and seen that, they'd be copping a lot of flak. [Alan]

Given the pressure and the inducements, youthful managers would appear almost inevitably to break company rules or the law, knowingly or not, in their directions to staff. But in setting up a system in this way, the blame is shifted downwards while the profits rise upwards.

What scope then is there in such a system for workers to resist managerial control, either individually or in a more organised fashion? Trade unions are forms of organised resistance. In view of what has been said above, it will come as no surprise to learn that there is very little unionisation of workers in either the fast food chains or corner shops. McDonald's discourages union activity in its stores, and the only fully unionised outlet in its 8000 stores is in Managua, Nicaragua (Labour Research, 1986). According to Alan, the company he worked for took a strong line: 'If anyone ever talked of forming a union, or some sort of group to represent the workers, straight away they'd get jumped on. You know, threatened with termination.'

With a young workforce unclear on its rights, the absence of countervailing power in the form of a union leaves the field free for autocratic management.

> I remember kicking up a stink. Cos after close you'd be left at the front at two in the morning . . . They really got shitty about me asking questions, and eventually I got the answer that the award, or somebody, said that they only had to provide transport when there was no transport available, and taxis were classified as public transport, we could get taxis. [Joyce]

The inadequacy of individual resistance in this setting is clear. If an employee complains about, for example, the lack of gloves to protect her or his hands in hot washing-up water when cleaning up at the end of the day, they would 'get laughed at' said Joyce. Alan added 'Who's he gonna complain to? Store manager? Store manager would say, "If you can't handle it, don't do closes", and then eventually they'd phase him out cos he is going to be a troublemaker. Just leave him off the roster.'

Acts of resistance reported as part of this survey were mopping the floor in an unapproved way, throwing food around, and taking more food than allowed for a meal. David said: 'If there had've been one manager caught you mucking around or not doing your work he would have just kicked you straight out.'

The final act of resistance is resignation. Alan lost two staff members a week: 'People waking up, weren't really getting paid for the work they were putting in.' Looking back, Joyce said: 'When I was there, I thought it was good. Cos I hadn't earned any money or anything . . . but looking back, how hard you worked and what you were doing, it was a pittance.'

Conclusion

Much of the fast food labour process can be understood in terms internal to the work organisation: wage coercion, detail division of labour, separation of conception from execution and the beginnings of mechanisation. However, the firm's externally oriented strategies (to product and labour markets) have also influenced their job and organisation designs. Keeping costs low for a mass market by employing teenage casuals, usually unskilled, inexperienced and often unpredictable because they are not yet used to alienated labour, has also brought in its train a detailed division of labour and a coercive and autocratic management practice. Emphasis is placed on a stringent system of rewards and punishments, on the consequences for rostering of failing to offer loyalty and commitment and on the hyperbole of teamwork, competition and bonuses.

While there are positive effects which flow from teenagers', especially students', experience of casual work in the way of increased feelings of independence and self-esteem (Coventry et al., 1984; Wilson et al., 1986), and in terms of learning to negotiate the culture of work (Reeders, 1986), there is cause for concern in the casual employment pattern emerging in post-war Australia. In some cases, the kind of work young people are exposed to resembles that of third-world 'export processing' plants where there are few actual rights and little protection by unions. These are secondary labour market jobs, offering little in the way of skills that have value outside the firm. Increasingly, full-time jobs are giving way to part-time or casual work which no longer offers a wage or a career suitable for adults or breadwinners. Many casual workers who are not also students (and this is especially true of teenage females) want to work longer hours but cannot find the opportunity (Macklin, 1982).

Returning to the question of the effects of so-called post-industrial trends, in the case of the fast food industry the evidence is clear. For the mass of workers, skilled satisfying work which meets the varied needs of customers is by design excluded from the labour process. What remains is alienated work in a smart uniform. The uniform, the advertising and the decor are all intended to obscure the fact that fast food is factory food, produced on factory principles and wholesaled direct to the customer.

That pressured and unskilled work is experienced as degrading or unsatisfactory work by those who do it in this industry is evident in the high rates of labour turnover. Young casuals are voting with their feet in large numbers. The question for those concerned about the rights of youth and the effects of their experience in such work is whether nonetheless they may come to accept this early experience as a standard by which to judge other work. Will they eventually come to expect the absence of unions, poor quality work, arbitrary management decisions, and the lack of any real chance to use the whole range of their initiative and competence?

11 Telephone exchange maintenance

LEN PALMER

In this chapter I want to show the relevance of labour process analysis to technological issues in Telecom Australia. Of course, there are many areas of work processes in Telecom, but my research is on the kind of work carried out by the skilled and semi-skilled workers who maintain telephone exchanges. Telephone exchanges today are automatic switching centres that direct telephone calls and data signals to their required destination. 'Data signals' are the raw information of computers when they are sent over a network. They can be sent over the same network as telephone calls, or over special networks set up for their use. Telecommunications and computers were once fairly separate, but are now converging extensively.

The main job of exchange maintenance staff is to keep the exchange running by analysing fault conditions, repairing faults and regularly testing the equipment to find faults and non-standard conditions that might affect service. The whole business of transmitting signals over landlines and through space (microwave and satellite) is called telecommunications. Theoretically, telecommunications are part of what Marx called the 'circulation process' of capital, as distinct from the 'valorisation process', where surplus value is produced. The extension of labour process theory to the sphere of circulation and the State is discussed in Palmer (1985; 1986b). See also Marx, 1967; Baxter, 1979; Frankel, 1983.

Telecom is not the only body that provides telecommunications in Australia (another example is the Aussat satellite corporation), but as a monopoly State corporation it is fundamental to the telecommunications industry. The extension of labour process theory to the sphere of circulation and State corporations is discussed elsewhere. (Palmer 1985; 1986b). Telecom was preceded by the Postmaster-General's Department (PMG) before 1975. In this paper I will use the term Telecom to cover both entities.

Many important changes have taken place in the area of telephone exchange maintenance over the last 25 years or so, with both the work and telephone exchange equipment changing in this time. Such processes of change have occurred not only in Australia but also on a worldwide basis, but different processes in other countries are the product of their different economic, political and historical contexts. Before describing the Australian changes, however, some definitions are necessary.

Most people use the term 'technology' to mean machinery. A major problem with this use is that it tends to disguise or forget the social basis of machinery design and introduction. Some traditions in industrial sociology that have been academically and popularly influential distinguish between the 'social' and the 'technical', and treat the 'technical' side of things as socially neutral; that is, they ignore the social processes that produce those technical developments (Braverman, 1974: 184–235; Thompson, 1983; Palmer, 1985: Appendix F). For present purposes I will not use the term 'technology', but refer simply to machines and equipment including whole systems of equipment.

Labour process analysis, as shown in Chapter 1, encompasses a way of theorising many issues relating to contemporary work. One important issue of debate concerns the 'division of labour' and its relationship to the machinery or equipment. The concept of division of labour has several meanings or uses and levels of application. The 'social division of labour' was defined by Marx to refer to the dividing up of social or societal tasks that any society deemed important (Marx, 1967: 331–4). An example of this might be the social arrangements that apportion activities to the footwear industry, the apparel industry, etc. Of course, in contemporary societies these 'arrangements' are huge socio-historical constructs: institutions, ideologies, ways of life, etc. But what was salient for Marx was that this was a universal phenomenon—all societies do it.

Of greater interest for Marx was the need to show that the division of labour took on a specific form in specific societies. He wanted to show that in capitalist societies we could talk of a specifically capitalist division of labour. For this he suggested a different concept of division of labour and, for capitalist societies like ours, he pointed to the historical development of a 'manufacturing, or detailed division of labour' (Marx, 1967: 318–47).

Braverman (1974) was also interested in the 'capitalist division of labour', and analysed the special changes to it following the detailed and long term studies of F. W. Taylor around the turn of this century. Without going here into Taylor's studies, and the changes to work practices that followed, Braverman was able to talk of the separation of 'conception and execution' which was intensified by the principles of what Taylor called 'scientific management' (1974: 113–18). The 'conception' side of this division refers to the power that owners and managers have by virtue of their ownership and/or control of factories, workshops, machines, etc. (known as the 'means of production') to control what is going on. Activities within this conception side of work include deciding what is to be produced or sold, the machinery used, the price and what the workers will do. In short, it includes all the things that are usually called management. This leaves the workers to 'execute' these decisions, effecttively reducing the control that workers have over their work. The division of conception and execution might be called the capitalist division of labour at a broad level.

My contention is that this separation of conception and execution applies no less in Telecom and telephone exchange maintenance than on the factory floors that Marx and Taylor analysed, and the offices that Braverman investigated. In Telecom this takes the general form that managers are occupied with conception activities, while technical workers are occupied with the execution of the rules, procedures and work practices that make up their working lives.

I have referred to the tendency in popular use, and in certain schools of industrial sociology, to treat the technical aspects of work and work organisations as neutral. A perspective that contains a different kind of problem, but is equally popular, is one that views machines (technology) as *causing* changes to work. This might be referred to as 'technological determination' (e.g. see Jones, 1982). Expressed in terms relevant to my research fieldwork, the following question might be asked: Does telephone exchange equipment dictate the work that maintenance workers do? Or, since my study has an historical dimension, Have changes to telephone exchange equipment caused changes to maintenance work? I should indicate that my initial 'hunch' in the research was that the equipment *was* causing changes to the nature of maintenance work which, as it turned out, was only partly correct.

Before attending to such an empirical question, one further distinction must be made. It is useful to separate the design of work from the design of equipment for conceptual purposes, even though they are obviously closely related to one another. Both are in the domain of management, and workers and their unions have little legitimacy in such areas usually referred to as 'management prerogatives'. I like to use the term 'design' to indicate that human 'designers' have been at work, to emphasise the *social* basis of the process.

Equipment design and work design roughly approximate the terms 'technical' and 'social' mentioned above. I want to argue that such distinctions must be used with caution because they are both the result of social processes. The design of work is what people usually mean when they use the term 'division of labour'. To distinguish this from the other concepts of division of labour mentioned above, I propose to call this the 'narrow' division of labour. Bearing this in mind, that management has control over both work and equipment, I will define them in the following way: equipment design refers to the way that machines or equipment operate, and the work that has to be performed by workers because of that design. In other words, machines dictate *what* workers do and to a certain extent *how* they do it: the materials, pace, tools, etc. *Work design* refers to the process whereby owners and managers decide *who* does the work, that is, which workers do the jobs involved, how many workers are employed, etc. (Palmer, 1985: 236–46).

To summarise, I have so far identified the following different concepts of division of labour; a 'social' division of labour common to all societies. In capitalist societies there is a 'broad' division of labour, what is also called the division of 'conception and execution', which gives managers control over machines and workers in capitalist societies, and a 'narrow' division of labour, here called the design of work, that governs which workers do what, under the control of managers in capitalist societies.

The study

In this study two main qualitative research methods were employed. The general field of ethnography (Hymes, 1974) theorises the role of interview and participant observation, as well as observation in grasping the important issues in any field of study. In this regard my background of twelve years as a technical worker made me an insider in more than the usual sense of participant observation. This experience also clearly gave me contacts and knowledge that I did not have to acquire when approaching the research as a researcher. On the other hand, as Williams (1981) discovered, the access to unions and workers placed limitations on access to management because of the suspicions of workers and their unions. This class dimension of research needs detailed theoretical identification in methodological teaching and writing.

The second research method I used was documentary analysis. Using primary sources and important secondary sources, I was able to gain access to management thinking 'in the broad' rather than the impressions of individuals, particularly over past events. Since much of what I studied was Telecom and union policy, the use of public (and occasionally secret) documents is essential in this kind of research.

A review of the changes in Telecom over the last 25 years reveals a marked trend towards centralisation in the provision of telephone services. In work organisations, centralisation generally means an increase of control over the various elements by management. Alternatively, decentralisation can also be a device for tighter control by central management, and perhaps the opposite of centralisation is autonomy. With Telecom this involved the centralisation of installation and maintenance of customers' services (in homes as well as in all sizes of businesses), as well as telephone exchanges. Both the design of exchange equipment and the design of work were centralised over this period.

The *equipment designs* that telephone exchanges use have to be common to the whole network, and these have been redesigned and replaced several times in the period I studied. Today they are highly centralised and based on computer-controlled switching (Palmer, 1985: Appendices A–D; Barr, 1985; Reinecke, 1985). I have indicated above that machines dictate *what* workers do, and to a certain extent *how* they do it—the materials, pace, tools, etc. Different machines require different associated activities and skills. For example, in modern automated nail production, there is no place for the skills of the blacksmith who once made nails at the forge. While it is true that these activities are dictated by the machine/equipment, it is the managers who have selected that machinery, in the light of the machines available, their aims on profitability, labour costs and productivity. This is an argument restored to the agenda of labour process studies by Braverman.

In the case of Telecom exchange equipment, the introduction of new equipment had the effect of changing what work was done by exchange maintenance workers. I will say more later about the way that the *introduction* of new machinery or equipment can be used by management to change work design under the guise of the 'demands' of the equipment. The design of all machinery or equipment is the result of a number of factors: what its designers intended it to do, the available technical means, and consultation with prospective buyers or a knowledge of their requirements. The way that exchange equipment operates determines just what tasks are related to its maintenance. The change to or introduction of various equipment systems by Telecom over the period studied achieved a number of aims: improved productivity through simpler maintenance effort, less maintenance costs, fewer maintenance workers and lower capital equipment costs (Palmer, 1985: Appendices). These changes reflect a view of what is important, of what is valued. No matter how much we may agree or disagree with these values, they represent values nonetheless. This is what is meant when labour process theorists like Braverman argue against the neutrality or value-freedom of machinery.

As shown in Chapter 1, the use of new equipment to reduce the skill levels of a group of workers for various reasons is a widespread problem for workers and their unions throughout the Western world. It is not easy, however, to argue that deskilling resulted from the introduction of *new equipment systems* in this case. In Telecom, different skills became necessary (e.g. less emphasis on mechanical skills and more emphasis on analytical skills) but I am not inclined to agree that an increase or decrease in skill has occurred; there has merely been a substitution of certain skills for others. A final point on the introduction of new telephone exchange equipment systems is that, for the most part, the unions, tenacious opponents of changes to the redesign of work, did not oppose the introduction of new equipment, but rather struggled over the terms and effects of its introduction.

Just as dramatic was the process of centralisation of *work design*. What makes this especially interesting is that this process was one of struggle between management and unions. This makes the first point about the case study. That is, the process of work redesign was one of political and industrial struggle over the degree of centralisation that would be introduced. The outcome of this process was the adoption of a less centralised work design than Telecom wanted, although more centralised than had been the case in 1960. The other side of this is that the unions, especially the Australian Telecommunications Employees Association (ATEA), succeeded in achieving a less centralised version of work design than Telecom wanted. What was remarkable was that many observers and most participants saw the process of work redesign as arising from the introduction of new *equipment systems*. This is an example of what I have referred to as 'technological determinism'; that is, the changes to work were seen to be caused by the new exchange equipment. Although the introduction of new equipment is linked with the redesign of work, the attempt to centralise work was separate from the changes to the equipment, even though Telecom always used the argument of the technical requirements of the equipment to insist on the necessity of changing the way exchange maintenance work should be done.

The history of the process of struggle over work design in Telecom Australia involves a bewildering array of technical terms, abbreviations, acronyms and jargon, some of which are necessary to fully understand what happened. Rather than enter into this 'world', I propose to offer here a brief and simplified account of the process of struggle over the extent of centralisation in the redesign of telephone exchange maintenance work. This account deals only with several of the most important issues central to studies of the labour process. An examination of the events in this process of struggle reveals that the issues which were constantly the basis of dispute between unions and management were telephone exchange closures, deskilling, job elimination and job satisfaction.

Telecom sought to centralise work by making the 'district' the basis of exchange maintenance. To explain the 'district' concept, it is important to understand that, until the early 1960s, the local telephone exchange was the basis of most aspects of telephone service. There were many exceptions to this arrangement, but this localised and relatively decentralised design was the general form. In fact, the telephone exchange was the base for a number of other aspects of telephone service, such as installation and maintenance of (small-scale) customers' equipment, as well as service complaints, etc.

The idea of the 'district' was to group a number of telephone exchanges together and to organise the work of those exchanges on the basis of the district rather than the individual exchange. In other words, the design of work was centralised to a greater degree than had existed before. The first stage in this process was to 'prune' some of the non-maintenance activities from the exchange, and to exercise greater control over the exchange staff. During the 1960s, installation and maintenance of customers' services were centralised to district centres and complaints services were centralised to regional and state centres.

Another aspect of the push for centralisation was the policy of exchange closures. In many districts a decision was taken by Telecom that some of the exchanges that had previously been staffed were to be closed and run on an unstaffed basis. This, as might be imagined, spurred a reaction from the technical workers involved. The ironical union joke at the time was that the new exchanges could be attended by a man and a dog, the dog to guard the exchange and the man to feed the dog! However, many workers at this stage were genuinely worried that the standard of service would deteriorate, and job satisfaction would deteriorate as the sense of belonging that came from the exchange-based design of work was lost because of the district-based design. As a result of the proposed amalgamation of exchanges into districts, a long struggle was begun over the size of these districts, because staffing levels (numbers and grades of exchange workers) were based on the amount of traffic the exchange carried, including the number of lines.

A very important aspect of the struggle over the redesign of work was deskilling. That is, not all workers would have the opportunity to perform the same degree of skilled work. To keep skills you must practise them but, with district-based maintenance, certain specialisations were proposed by Telecom that would reduce the range of skills practised by individuals. New divisions among workers were likely, as some workers were to be allocated extremely simple repair tasks, for example, replacement of printed circuit boards, while some were to become specialist 'troubleshooters'.

Job elimination is often an important consequence of new machine introduction and work redesign and, as the general explanatory framework of labour process studies implies, this can be a direct intention of managers and owners (Kelly, 1982). Job elimination differs from sackings or redundancy in that new jobs are not created and existing jobs are not retained. Job elimination in Telecom was a direct result of job redesign and equipment redesign. Fewer workers were needed if hitherto staffed exchanges were closed. Repair and other maintenance work was provided from district centres when necessary. By holding the workforce levels steady while increasing the productivity of the existing workers, Telecom aimed, in 1973, at eliminating 1800 jobs by 1980 that might otherwise have been needed. These included apprenticeships, traineeships and semi-skilled and skilled jobs that simply did not come into existence (Palmer, 1985: 39).

In empirical research, it is often important to look at some of the crucial moments of historical processes to illustrate the kind of relations-in-struggle that make up those processes. An important watershed occurred in 1978–82, when a number of the issues that I have mentioned so far came to a head. Very briefly, the events began with a bitter four-week industrial dispute involving the then Prime Minister, Malcolm Fraser, Cabinet, and the Arbitration Commission. Union bans were placed on all maintenance and customer installations and mass stand-downs of approximately 4000 workers by August 1978 were instituted by Telecom (Palmer, 1985: 61). Intervention by the ACTU through its President Bob Hawke finally achieved agreement on a nine-point plan.

One crucial breakthrough was the trial and independent evaluation of the two separate work designs for telephone exchange maintenance proposed by Telecom and the ATEA. After these one-year trials Telecom accepted a version of the union-designed maintenance workplan in what is widely regarded as a world first in the introduction of new equipment and work design. In addition, an important industrial benchmark was established, a 'memorandum of understandings' covering the agreed design and future consultation prior to the introduction of new equipment (Telecom, 1982; Mathews, 1985a: 70–3; Reinecke, 1983; 1985; Barr, 1985).

Several things should be emphasised about this watershed period. Firstly, the union's design for exchange maintenance work was compatible with exactly the same Swedish exchange equipment as the design pursued by Telecom. Secondly, the settlement agreed on a work design that was significantly less centralised than the one which Telecom had planned. Conversely, by 1982 Telecom had achieved a considerable degree of centralisation compared to the beginning of the 1960s. Finally, it is important to understand this settlement as part of a much longer process, as being the accumulation of issues over a longer period.

A further, even more significant, conclusion can be drawn from the trade-off between Telecom and ATEA. The two designs reflected very different perceptions of what was important, and embodied the different perceptions and values. To make this clear, I will briefly describe the alternative designs and sketch the process of their emergence.

Telecom's design for exchange maintenance in 1976 was based on what it called exchange maintenance centres (EMCs). These EMCs were district centres from which all exchange maintenance was planned and performed for that exchange district. Despite the long process of resistance of the centralisation process by unions and workers, the EMC design constituted an increase in centralisation over prior maintenance work plans (Palmer, 1985: Appendix E). The design consisted of an hierarchical four-tiered system of technical support centres at the national, regional and district levels, and at on-site work.

The national support centre was to control all software associated with the computer-controlled national network. The regional centre was to operate at state and country levels in control of district EMCs, and more limited activities associated with customer data. The EMCs had the role of overseeing the performance of exchanges in their district, and of analysing faults and operating data, as well as fault repair.

The 'on-site work' proposed by Telecom consisted of the worst possible arrangements imaginable from the point of view of the maintenance workers concerned. This work would consist, according to the union, of '. . . work on the MDF [Main Distributing Frame, boring, low-skill wiring activities], and fault rectification by card [printed circuit board] changing techniques. This latter function would be at the direction of specialist staff [from the district centre].' (ATEA, 1976: 247). It can be seen from just these scant details that, for all except the elite few who worked in the upper reaches of the hierarchy, a deskilled and divided future would face most exchange maintenance workers.

The system as a whole raises some of the issues I referred to earlier. For example, the kind of work found at the national level is different to the work at other levels of this maintenance hierarchy. The kinds of work tasks found at the upper level involve abstract and analytical skills, as well as thinking through logical problems and solutions. There is also more responsibility and pressure (and prestige) here than at lower levels of the system, with the national network to look after. The hierarchical division of the work reinforces divisions between the workers at the various levels of the system, and fosters competition between differently graded workers (with different wage rates), and promotes lack of sympathy with, and understanding of, the different tasks involved.

The response by unionists to these plans quickly gained momentum and talks between Telecom and ATEA broke down in late 1977. After Arbitration Commission intervention, the ATEA won an agreement to propose an alternative work design for exchange maintenance, based on what they called exchange support centres (ESCs).

The ESC design was based on a self-conscious attempt by the union to oppose Telecom's heavily centralised approach and insert 'retention of responsibility in Telephone Exchanges' (ATEA, 1977: 246). The ESC was still to be a district centre but its role was different. Instead of having a controlling function, it was, as the name suggests, to be a *support* centre, operating as a technical back-up and advice resource to the exchanges in that district. This was acknowledged as necessary, especially during periods of new equipment installation and modification. It was also necessary because some faults are relatively infrequent, especially in the computer processors that control the switching. As I suggested before, skills have to be used to be retained. To overcome the tendency to create 'experts' through practise, the union advocated the constant rotation of jobs throughout the ESC system.

I mentioned earlier that the 1978–82 dispute involved a trial of the two maintenance designs, Telecom's EMC and ATEA's ESC. The trial ran between July 1979 and June 1980. Under the auspices of the Arbitration Commission, independent observers performed evaluations on each design during the year-long trial, finding the union work design preferable when judged according to the criteria of efficiency, standard of service available to the public, job satisfaction, career opportunities, maintenance of technical standards and expertise and the public interest in general (ATEA, 1982).

It is clear that the agreement ratified by Commissioner Clarkson in 1982 was a compromise between the systems proposed by Telecom and ATEA. The new maintenance structure was to consist of the following elements: a national support centre (NSC), a state support centre (SSC), a district support centre (DSC) and, significantly, the exchange maintenance group (EMG). This design is the basis of metropolitan, country and specialised exchange maintenance today. The differences from Telecom's desired design become clear when the working details are described.

The national (NSC) and state (SSC) levels of the system are similar to Telecom's EMC design described earlier. The big differences relate to the district and exchange levels of the system. The District Support Centre (DSC) *assists* local exchanges with failures, so many more local exchanges are staffed than would have been under the scheme that Telecom planned for. The local exchange maintenance workers (EMG) treat their local exchange as 'their own' (an important aspect of job satisfaction as noted above). Job rotations, especially with the DSC, reinforce skill retention and resist deskilling of the exchange staff.

More importantly, the ways in which the different centres in the union work design relate to one another tell a different story to that of the system that Telecom pursued. This can be best illustrated by recounting the procedure for dealing with a serious exchange fault. The local EMG advises the district centre of a serious fault occurrence, but can proceed with analysis and repair on its own. If that repair is not effected in 30 minutes, the district support centre (DSC) takes over. The EMG can call in the DSC at any time, for example if the fault condition is beyond the experience of the local exchange staff. For its part, the DSC must pass control of such a situation after another 30 minutes to the state centre (SSC).

Here we can see an innovative response to the problems of equipment sensitivity and standards of service on the one hand, and the retention of skill, career prospects, job retention and job satisfaction on the other.

Conclusion

I have shown that the process of centralisation of telephone exchange maintenance in Australia occurred as a result of the pursuit of work redesign by Telecom management. Had Telecom achieved the degree of centralisation it wanted, the technical workforce that maintained its exchanges would have been deskilled and lost jobs and job satisfaction. In other words a simple deskilling outcome has not eventuated in this case because of the industrial resistance and struggle that the union and its workers offered. Instead, some reskilling and job elimination was found. I have also shown that the redesign of work is not simply the result of the changes to the equipment. In the terms of the survey offered in Chapter 1, this demonstrates the 'social shaping of technology' with all the historical detail of the case study approach.

Central to this account is the fact that the process of equipment redesign was pursued by Telecom quite separately to the struggle over work redesign, and that all three work designs (EMC, ESC, and EMG) were feasible and practical in relation to the equipment concerned. This emphatically refutes a technological determinist explanation of changes to telephone exchange maintenance in Australia. In other words, the political and industrial outcome was not dictated by the equipment employed. This finding is at odds with the terms of the debates of the day and the records of the process found in public documents. More importantly, this finding raises serious questions about the easy assumption of technological determinism in popular consciousness.

Conversely, the equipment systems employed could support different work designs, reflecting different perceptions of what was needed, and what was deemed important or valued. This is a vital conclusion to be derived from such research as this. This means that a number of work design options exist, or are potentially feasible, for a given equipment system. The actual options that are implicit in any given equipment or machine system depend on the desire or values that are operating, not the technical imperatives of the machinery. This places decisions about the effect of the introduction of new equipment firmly back into the social and political spheres. In the normal run of things in institutional settings like Telecom, the 'structural feature of control' identified in Chapter 1 as symptomatic of the 'sociology of work' approach is found in the relationships between management and technical workers. In the normal run of things, management decides the options that will be taken up.

Earlier in this chapter I showed that the introduction of new exchange equipment performed certain tasks for Telecom management and thereby reflected certain value orientations. So how can the equipment be said to embody social values on the one hand, yet support three different sets of value perspectives on the other hand? The solution to this question is in the division of labour. The 'narrow division of labour' (design of work or division of execution) did change in content as a result of struggle, and indeed it is in this area that a significant challenge to the nature of social relations was located. However, control over who does the work and which workers do what is still retained by Telecom managers and given by the particular (broad) division of labour typical of capitalist societies. Even though the exchange equipment could support three different work designs, crucial features about the division of labour did not change. The division of conception and execution, between managers and workers remained. It is true that, *temporarily*, the right of management to conception was challenged, albeit through the highly mediated processes in the State institutions of the arbitration system. But the social relations described by the division of conception and execution, the 'broad division of labour', remain intact.

Further, the division of conception and execution also gives management the right to design (or at least select from the available design options) and introduce equipment. It is salutary in this case study that Telecom's control over equipment design was not challenged by workers throughout the 25-year period studied. In fact, the ATEA accepted the introduction of several generations of new exchange equipment systems, but sought rather to struggle over the effects. The division of labour remains a capitalist one.

My account of the changes in Telecom Australia so far has used the distinction between the design of work and the design of equipment. I indicated when first introducing those terms that they were 'obviously closely related to one another'. It should be clear from the analysis that work design and equipment design should not be distinguished on the basis that one is a 'social' process and the other is not. In Australian telephone exchange maintenance one process (work design) was struggled over, and the other (equipment design) was not. Both processes expressed varying social intentions, perceptions and values.

We can focus attention now on the *relationship* between the equipment and the work to which it gives rise on the one hand, and work design on the other. One dimension of this relationship is that they are both social. Another aspect is that they both express social intentions and values. A further dimension, mentioned earlier, is that the *introduction* of new equipment systems can be the opportunity for management to change work design, even if this is not necessitated by the equipment.

This last aspect is especially important, because it reminds us that management knows about the *existing* work design, and perceives its strengths and shortcomings from a management point of view. This suggests that the relationship between the machine, or equipment system, and work design is always relative to an existing work design (division of execution). This is true whether management is contemplating changes to the design of work or to the design of equipment. This relationship holds as well to what I have called the 'broad division of labour'. That is, the relationship between the design of equipment and the design of work is also relative to the division of conception and execution, even though this is rarely challenged and never permanently changed in capitalist societies such as our own.

This permits me to attempt a definition of technology for the first time. In Chapter 1, the 'social shaping of technology' was made explicit. Here some detail can be specified. I think we should define 'technology' in relation to both equipment design and work design. This means that the term 'technology' implies a division of labour as well as machines, and that it implies a given division of labour in its 'broad' and 'narrow' meanings. That is, the term 'technology' implies a given (existing or contemplated) division of conception and execution, and a given division of execution. By this definition, when we use the term 'technology' we will not merely mean machines, in the technological determinist fashion that is usually implied, but use it in a way that shows the how the specific capitalist social relations in the division of labour shape what technical decisions are made, and the implications for workers. This amounts to a politicisation of technology that opens up debate about the way in which work is designed, and about the way in which equipment is designed and introduced. As Mathews (1985: 108–26) has recognised, the introduction of new technology represents an *opportunity* for workers and unions to fight to insert alternative work and equipment designs, derived from alternative social values and perspectives.

12 Trade unions and work organisation

JOHN MATHEWS

As the studies in this book demonstrate, and as the wider post-Braverman literature confirms, the workplace is the site of a major struggle, not just over wages and conditions, but over the forms and organisation that characterise the process of work. This has quite fundamental industrial, economic and political ramifications. Both management and trade unions are drawing important conclusions from labour process studies, and are developing sophisticated policies and negotiating frameworks that underlie current efforts to promote industrial restructuring and manufacturing technological development.

Technological change is now widely understood to be a *social* process, evolving not in a predetermined manner, but as the outcome of a series of negotiations, struggles and concessions between groups in the community (MacKenzie and Wajcman, 1985; Mathews, 1985a; 1986). Such a statement opposes, on the one hand, the 'technological determinist' view that change occurs in a unilinear direction under the influence of purely objective factors such as efficiency and productivity; on the other hand it opposes the conventional neoclassical economic view that change is influenced solely by profit and loss calculations and adjustment of factors of production. These are indeed blinkered ways of analysing changes in the productive base. A perspective that focuses on the *choices* involved between different options at any point of technological change, and on the interests of the parties involved (such as employers and unions), generates quite a different and illuminating reading of the direction taken by technological change since the Industrial Revolution—and provides us with a handle on the present changes and the directions in which they *ought* to evolve.

Numerous examples present themselves. In the early cotton mills a type of automation (the 'self-acting' mule) was sought to curb the power of the organised spinning workers, by putting them out of work. There is now evidence available that demonstrates the explicit social goals of the promoters of this technical advance—and how the effects of the self-acting mule in practice were not quite as intended by the promoters, due to counter-initiatives by the groups of workers involved. These struggles led the cotton industry to take quite different trajectories in the UK and USA (Lazonick, 1979; 1981; Bruland, 1982). Much more recently, management control over machine tooling, a traditional preserve of craftsmen, was enhanced by the introduction of numerical control in the 1950s. This is discussed further below (Noble, 1979; 1984). This case is complicated by a further development too, namely the use of electronic microchips in computer numerical control, which has given operating workers much greater potential to regain a measure of control over the work process (Shaiken, 1984; Shaiken et al., 1986). The struggles are not always between workers and employers, but frequently occur between one group of workers and another. For example, print workers in the nineteenth century were able to utilise the innovation of the Linotype machine to oust women workers from typesetting (Cockburn, 1981).

Through these and many more examples, we are coming to understand the far-reaching *social dimensions* that characterise the process of technological change. This is still a minority view; as Badham says:

> Quite often discussions of technology remain dominated by outdated and simplistic critiques of technological determinism . . . simplistic pigeon-holing of theorists into technological 'pessimists' or 'optimists' . . . (Badham, 1986: 288).

Badham himself argues for an explicitly 'techno-choice' approach to the analysis of technological change—an approach which facilitates the development of a framework needed for the public control of technology. The conceptual difficulty involved in taking this step is that choices between technologies are not taken in the abstract, but in the concrete conditions offered by each; the options are constrained by the choices made previously. The choice between steam-driven turbines and petrol-driven engines to power motor vehicles, taken early in this century, clearly limited the possibilities for changing direction later, even now, when the pollution problems of vehicle exhausts have become a major concern.

In this way, technology can be understood as the aggregate of a series of choices made in the past; it carries the 'scars' of past struggles and compromises. If we define technology broadly enough, as encompassing not just the hardware of machines and tools, but also the software of programs, routines, and decision structures, then the process of technological change can be seen to subsume many of the changes that we already understand as being political; from this perspective, the social character of technological change is manifest.

This chapter looks at the way in which the study of job design has revealed new layers and dimensions to the process through which the subordination of workers to capital is sought. It also explores how the technology of the workplace has become inextricably involved in this process, from the introduction of the moving assembly line and the development of 'time and motion' study to the present reorganisation associated with the applications of micro-electronics.

It could now be argued that a process of *democratisation of technology* is under way, and that this is manifesting itself most clearly through collective worker intervention in the workplace. Unions are beginning to transcend their traditional hostility and opposition to technological change in favour of a process of negotiation and positing of alternatives or imposing new criteria (Frenkel, 1987). This chapter is devoted to the exploration of this new species of trade unionism.

I shall first discuss the protective measures that need to be in place before workers and their unions can take a constructive attitude to technological change, then go on to analyse the elements of a labour process *policy* that could guide the negotiations of unions with employers over changes in work organisation. Out of a host of possible issues, I shall focus on five: questions of skill, job content, divisive job measurement and wage systems, overcoming hierarchies of control, and developing the criteria needed to evaluate proposed technological changes and indeed to initiate and propose changes.

Protective responses

Of course, the unions and their members have suffered grievously from the effects of technological change in the past. Numerous studies have demonstrated the immediate employment effects of unregulated technical innovation that has thrown workers out of a job, or rendered the skills of whole groups of workers useless (ASTEC, 1985; Senate Standing Committee on Science, Technology and the Environment, 1987). For the workers involved, the proposition that a dynamic economy might offer some of them alternative work at some time in the future is a dubious consolation. It is intolerable that the burden of technological change and industrial restructuring should be borne entirely by those who are so brutally displaced.

There have been bitter lessons, bought dearly. They have resulted in unions throughout industry confronting employers, when technical changes are proposed, with demands for job security and protection, and for retraining of displaced workers. These demands force a degree of social responsibility on the employers, and restrict their ability to discard an unwanted workforce at will. In Australia, employment protection standards have lagged behind those secured in many other countries. The long-awaited decision of the Australian Conciliation and Arbitration Commission (1984) in the Termination, Change and Redundancy case, handed down in two instalments in 1984, moves some way towards consolidating the position achieved by some unions and brings Australian employers more into line with their overseas competitors. Since then, the integration of this decision into individual awards has been slow.

Some unions have been able to move beyond mere job protection, and have negotiated technological change agreements with employers. These are now common in the Scandinavian countries, in northern Europe and the UK, and they have been won in Australia as well, notably within Telecom in 1980. This agreement flowed from a celebrated dispute, discussed in Chapter 11, between Telecom and the telecommunications engineers over the introduction of computer-controlled switching apparatus. The engineers, through their union (the Australian Telecommunications Employees' Association), agreed that the changes were desirable, but they insisted that the new technology did not have to be introduced in a way that would create an elite corps of maintenance staff. They successfully imposed their own job design proposals to accompany the new technology, preserving skills and career structure, and enhancing the efficiency of the whole telecommunications network (Robson, 1980). In line with the formal agreement, changes in technology within Telecom are carried through now only after specific written agreement on job specifications, skill levels, training and promotion matters has been obtained.

These technological change agreements impose a certain framework of consultation on the process of change and are therefore a step in the right direction. But before workers and their unions can drop their hostile and defensive attitude to technical innovations (an attitude wholly justified by the costs borne by workers over a century of unregulated change in Australasia), certain community steps are needed. It will not always be possible to protect the jobs of those workers whose skills are displaced by change; in these cases, the community should be generous in seeing that the costs of change do not fall disproportionately on the shoulders of those workers and their families. At the very least, financial compensation is necessary to tide the group over while alternative employment and living arrangements are made. A security net along these lines was recommended by the Myers Committee in Australia (Committee of Inquiry on Technological Change in Australia, 1980), but there are no current steps being taken to establish it. The community should also be generous in providing adequate training and retraining facilities for such displaced workers, as recommended in Australia by the Kirby Committee on labour market programs (Kirby, 1985). But, again, progress in implementing these recommendations is slow. In my view, until there is a humane and generous community response to the displaced victims of technological change, with steps being taken to cushion the social costs and effects of change before that change occurs, it is unrealistic to expect the unions to markedly change their role as defenders of the *status quo*.

Towards a labour process policy

Despite the unfavourable social environment in which technological change takes place, unions have been able to make some progress in dealing with more than the *external* features of the proposed change (numbers of workers, training details, work value considerations). Unions and employers now negotiate a framework of participation and consultation, indeed of joint determination, in which the *internal* details of any proposed technical change may be settled before the change is carried through. These details will include the precise technical specifications of the proposed change, the manner of organising work, the job content and skill structure of the work, the different technical options available and other factors.

Not so long ago, and certainly into the early years of the twentieth century, these matters were normally the province of the worker and not of the employer (except in high technology process industries like the chemical industry). Yet that time now seems far away. Following the inspiration of Frederick Winslow Taylor and his system of 'scientific management', the employers staged what amounts to a pre-emptive coup at the turn of the century, and expropriated the workers' former control over the details of the labour process. In place of this direct control there was erected an immense management apparatus whose role was to issue precise work instructions and monitor every aspect of the production process with a chain of paper moving from desk to desk, and now from computer to computer (Cooley, 1984).

It is this control over the internal details of the labour process, over the content of work, that is beginning to engage the unions now, according to recent strategy documents (ACTU, 1987). In formulating this approach, they are claiming back no more than was lost a few short decades ago—but in the context of advanced computerised technology, rather than in that of the lost tools and trades of the nineteenth century. Unions are not attempting to roll back the workplace and technology to the turn of the century, but are claiming the latest technical advances and proposing new technical advances which place a premium on workers' skills, which provide a flow of current information allowing workers to take meaningful decisions, and which put an end to the slavish, authoritarian hierarchies of work processes modelled on the assembly line.

As mentioned above, we shall focus on the following five issues in a *labour process policy*: skill, job content, divisive job measurement and wage systems, job hierarchies and evaluation of new technologies.

Skill

The fundamental 'property' of workers is their skills, and the protection and enhancement of these skills is coming to be the central object of trade unions. Of course, in a sense this is already the case, for unions were formed of groups of workers who defined themselves in terms of their special skills, and they took steps, such as restricted entry to trades, to ensure that these skills were not diluted. This is a defensive approach to skill protection, and it can lead unions into confrontation with each other. A classic instance of this occurred in January 1985 in New South Wales between train drivers, members of the Australian Federated Union of Locomotive Enginemen (AFULE), and guards, members of the Australian Railways Union (ARU). The confrontation, leading to strikes on Hunter Valley coal trains, arose from a 26-month-old crewing dispute hinging on the introduction of guards into drivers' cabins without them necessarily going through the full training undergone by a driver. AFULE members were defending their notion of skill—but in a way which clearly cannot be generalised to the rest of the union movement.

I would submit that the way for unions to protect and enhance skills is to make them the centrepiece of negotiations over wages systems, job classifications, work organisation and technological change. Levels of skill should be referred to explicitly in job classification systems as spelt out in awards, and these defined levels could be referred to repeatedly in all subsequent agreements. Any proposed technological change would then clearly disturb these definitions of skill and classifications, and so skill would be the starting point of negotiations over the proposed change. This approach to skill is already implied by unions taking 'work value' cases to the Conciliation and Arbitration Commission, and seeking to have the remuneration of a worker or group of workers improved by appealing to notions of skill, experience and responsibility. The problem is that, in work value cases, it is not the skill which is the central object of concern, but the remuneration. This becomes clear when unions acquiesce to technological change which destroys the skills of many workers but enhances the skills of a few, and then argue a work value case to increase the remuneration of the privileged few. This is a use of work value cases that comes close in principle to the past pursuit by unions of 'danger-money' for workers exposed to excessively hazardous conditions. Just as unions have now learned that it is better to negotiate the removal of the hazards than to accept danger money, so unions will have to learn that it is better to protect skills directly than to trade them off against 'work value' considerations.

In this approach, skills are not defined in abstract terms (as 'human capital' or whatever) but in the practical terms of training undergone and ability to do the job (measured by proficiency tests). The professions certainly pay attention to these issues, and base their whole prestige and bargaining strength on the possession by their members of degrees, diplomas and certificates of one kind or another (Willis, 1983). Unions too have been mindful of the importance of various trade qualifications but, if skill is to become the central issue that it deserves to be, then unions will have to embrace a concept of lifetime training, constant acquisition and reacquisition of skills, certification of their members and restriction of jobs to certain levels of training. The notion foisted on us by Taylorisation, of jobs open to all because of the minimal level of skill required to do them, will have to be replaced by a notion of multi-faceted jobs requiring a high level of skill, judgement and responsibility, and open only to sufficiently trained workers (subject to a negotiable degree of employer obligation to provide on-the-job training).

This approach to skill is informed by the labour process literature and its discussion of deskilling (whether actual or potential), but seeks to make skill a central subject of negotiation in the concrete circumstances of a particular technological change. It is at the point of change that skills need to be defined most urgently. It is a practical proposition for skills to be defined and logged by unions, with the assistance of government, and to be built into models of technological change. This approach will have some immediate effects on unions themselves, leading them to adopt a more flexible approach to traditional categories like 'skilled' and 'unskilled' (normally defined by apprenticeship provisions), and to abandon obstructive demarcations. It will have wider effects on the community in the demands made on the educational system and the social process of skills acquisition. It is notable that industry, too, is coming to recognise this central importance of skills formation, and of its own role in developing skills—an approach which directly contradicts the earlier concern with fragmentation of skills via Taylorisation (Svedberg, 1986).

Job content

Reintegration of the labour process is now a trade union goal. Braverman has described how successive divisions of labour have fragmented jobs into meaningless 'detail' operations devoid of skill or interest. The major contribution of Taylorist 'scientific management' was to effect the ultimate division, between mental and manual labour, making the former the province exclusively of a management apparatus erected above the job-performing workers. Henceforth workers would not be 'paid to think' but simply to carry out a series of operations (or the same operation over and over again) and instructed. Workers have rebelled against this dehumanisation of work with a variety of techniques, such as sabotage and 'work to rule' (a tactic that makes sense only when work is done in accordance with rules rather than skills). These are tactics born of frustration, however, and are not the constructive tactics that will lead to a new social order.

As a general principle, when new technology is being introduced, unions should insist that the jobs created contain elements of both manual and mental labour, so that they entail both conception and execution of tasks. This is not a utopian demand, and anything short of it will see a perpetuation of the mutilating divorce between conception and execution that was accomplished by the Taylorisation of work. It is of great interest that the current wave of computerisation provides the opportunity for just such a reintegration of work, provided unions press for it and are armed with the appropriate policy and expertise. Let us see what this counter-thrust might be in the case of computerised machine tools, such as metal-cutting equipment. The introduction of microprocessors into machine tools is creating new possibilities for workers to regain control of their tools. These tools are called computer numerical control (CNC) systems. The microprocessors allow the machine itself to store programming instructions for several different cutting sequences, allowing them to be edited together on the spot. Depending on the power of programming, quite complex contours can be synthesised by a machinist, either through a keyboard (manual data input (MDI)) or by moving the machine itself to make a first component, and allowing the microprocessor memory to store the movements in digital form. This brings automatic machine tools through an astonishing complete circle for, as David Noble has shown so convincingly, the earliest concept of such systems was that of 'record-playback' (using magnetic tapes instead of the modern microprocessor) but it was consciously abandoned by the industry in favour of step-by-step instruction tapes. Noble argues that this was a conscious social choice, for 'record-playback' allowed workers to control the process, while NC passed control to engineers (Noble, 1979; 1984). Now computer numerical control (CNC) is bringing the potential for worker control, via programming, back on to the agenda. However, this potential is not always acted on. Noble cites the General Electric (GE) plant at Lynn, Massachusetts, as a typical example. Here machine operators are not permitted to edit programs, much less make their own, on the new CNC machines; quite often the controls are locked. Only supervisory staff and professional programmers are allowed to edit the programs. Noble (1979: 47) points out that the same type of technology is used with a very different organisation of work in Norway:

> The situation is quite different in the State-owned weapons factory in Königsberg, Norway, a plant with roughly the same number of employees, a similar line of products (aircraft parts and turbines), a similar mix of commercial and military customers, and, most important, the same types of CNC machinery (although here they tend to be European made rather than Japanese) as at GE. But in Norway the operators routinely do all of the editing, according to their own criteria of safety, efficiency, quality, and convenience; they change the sequence of operations, add or subtract operations, and sometimes alter the entire structure of the program to suit themselves. When they are satisfied with a program and have finished producing a batch of parts, they press a button to generate a corrected tape which, after being approved by a programmer, is put into the library for permanent storage.

Noble is making a fundamental point here. The very same technology is being used in one plant to frustrate the initiative of shopfloor workers, while in another plant the potential of technology to liberate and enhance workers' initiative is being realised.

There could not be a clearer example of the fact that work organisation is not determined by the technology employed. There is scope for collective choice over how the work will be organised and performed; and it is the task of the union to intervene to ensure that this scope is realised. It is through insisting on one form of work organisation, and militantly opposing the other (because it is fundamentally anti-human), that the union can contribute to a re-integration of the mental and manual facets of work, and thereby pave the way for genuine workers' control.

Divisive job measurement and wage sytems

Systematic job measurement and associated incentive payment systems were introduced in the 1880s as a conscious employer tactic to try to wrest control of the labour process from skilled workers organised in a sub-contracting pyramid (Clawson, 1980). The sub-contracting system had been a useful technique in the earlier part of the nineteenth century, when not many employers had the technical knowledge to run a factory and merely contracted to have the different stages done under the control of a skilled contractor. But by the end of the century it was being actively phased out in favour of direct employer control, using the carrot of payment by results (PBR), involving piecework and premium bonus systems (Littler, 1982). This system of control was in turn made obsolete by Taylorisation and 'scientific management', and yet it lingers on in parts of industry, particularly in the clothing sector, and is actually making a comeback in the clerical sector as employers drive data processing operators and keyboard operators harder and harder with bonus systems backed by electronic monitoring of individual performance levels.

Unions in Australia have made a stand against these divisive management systems. In line with ACTU health and safety policy regarding screen-based equipment, they have negotiated agreements covering such operators which outlaw electronic monitoring and do away with bonus systems of payment. The immediate objective of such clauses is to prevent operators from succumbing to repetitive strain injuries, but they have the wider goal of humanising work and promoting a group rather than an individual mentality on the job.

At a deeper level, incentive payments systems promote a narrow, instrumental approach to work; they focus on a simple, quantitative measure of tasks performed, and thereby militate against a more flexible approach based on skill and worker responsibility. The payments system goes hand in hand with the aim of monitoring worker performance down to the last detail, an approach which can be amplified with the introduction of computerisation. This is recognised in the trade union policies on screen-based equipment, and is described vividly in the context of metal machining by Shaiken et al. (1986).

Job hierarchies

Hierarchies of control at work are not 'technically necessary', although they are certainly embodied in many technologies, such as in the assembly line. Hierarchies of control, patterned on a military chain of command (and designed to secure the same unquestioning obedience by the rank and file of workers) were introduced along with factory production. They were imposed as a quite conscious choice by employers, in order to secure work discipline and to crush the spirit of independence that early factory workers brought with them from their previous modes of employment, as in domestic production (Thompson, 1968). Later, mechanisation was able to impose a further degree of discipline; again, not through any 'technical necessity', but because employers chose to develop a form of self-acting and self-regulating type of machinery that was experienced as an external disciplining force by workers, rather than as an adjunct to their powers. This line of development found its ultimate expression in the assembly line.

Unions now need to tackle the issue of authoritarian and hierarchical job organisation and control at two levels: the organisational and the technical. At the organisational level, the modern hierarchy of control using technical supervisors and forepersons accountable solely to the employer is a relatively recent management innovation. Unions need to oppose it, by counterposing 'work co-ordinators' who are elected from the workforce and are accountable to that workforce, rather than to management. Clearly there must be some form of co-ordination of work, but as long as the 'co-ordinator' is appointed solely by management and is accountable only to management, there will be no end to job hierarchies and no possibility of genuine industrial democracy. This is not to imply any personal criticism of supervisors in industry at the moment, nor to query the activities of the unions which organise them. Supervisors are workers too, but they are caught in a hierarchy over which they have no control. They will eventually share in the benefits which accrue to all workers by the elimination of such hierarchies.

Hierarchies and authoritarian work structures are also imposed through particular types of technology which treat the worker as an automaton simply receiving orders from a machine. Such technologies should be vigorously opposed by unions, particularly at the point where new systems are being installed. In particular, unions should invariably oppose the introduction of new 'assembly line' type technology which recurrently presents the job to the worker at a predetermined pace. This raises the issue of how unions should *evaluate* new technologies, and impose their own criteria of acceptability.

Evaluation of new technologies

This 'labour process' policy and the wider issues related to it are best implemented through the evaluation of proposed new technologies according to certain criteria. These criteria might include the following:

1. The technology should require specified skills which, in aggregate, are superior to existing skill requirements.
2. It should provide a well-defined career path for the workers involved with it.
3. It should provide a minimum of job classifications with uniform skill gradation between them (rather than highly skilled jobs for the few, and unskilled jobs for the many).
4. It should be capable of being used by workers without regard to gender, race, age or common disabilities.
5. It should not be authoritarian in its operation. For example, it should not pace workers nor dictate the pace of work; it should interact with the worker, offering options rather than predetermined outcomes. The more scope for decision-making given to the worker, the less authoritarian the technology (provided the worker is not stressed by being forced to take decisions without adequate information or without being able to foresee their consequences).
6. It should not fragment a task into a number of details. Rather, as an optimum, it should provide a clear sequence of operations leading to a finished product or activity, to be accomplished by a single worker. This criterion places a union-imposed limit to the operation of the Babbage principle.
7. It should involve both mental (planning) work and manual (execution) work, thereby reuniting the phases of conception and execution that were split by the Taylorisation of work. This involves building on, and enhancing, the worker's skills, rather than bypassing them.

8 It should not involve repetitive action. In particular, unions should insist that any cycle or task should take at least ten minutes to perform.
9 It should provide adequate stimulus, interest and variety.
10 It should favour group work targets, and not depend on monitoring of individual workers' performance.
11 It should in principle be comprehensible to all workers involved with it.
12 It should be without uncontrollable risks to safety and health.
13 It should provide a role for an elected work co-ordinator.
14 It should be socially useful.

These criteria can be used to evaluate new systems of work and indeed to act as specifications for the *design* of new 'humanist' work systems, in the same way that more conventional specifications of tolerances, efficiency, energy consumption and other issues are given to designers.

Practical implementation: opportunities and obstacles

Armed with a 'labour process' policy as outlined, unions are in a position to negotiate work organisation agreements and consultative frameworks that place the issue of co-determination of workplace issues firmly on the political and industrial agenda. This is a strategy for achieving industrial democracy through confronting concrete issues and solving them one at a time, in stark contrast to the fruitless efforts of the 1970s when 'industrial democracy' was sought through abstract demands (representative or direct democracy) and slogans that frequently bypassed the unions altogether.

What is the evidence that unions are prepared to embark on this process that will clearly take them well beyond previously narrow interpretations of the nature of trade unionism and of 'industrial relations'? Let us look briefly at one concrete issue that is of immediate relevance to workers, namely their health and safety at work, and see how unions have made some progress in tackling this as a *co-determination issue*. Unions have made great progress in negotiating a consultative framework at workplace level within which health and safety issues may be addressed. Health and safety agreements now cover a considerable part of Australian government employment, of Victorian government employment in both the public service and instrumentalities, and of the private sector, notably the airlines and the national pulp and paper industry. These agreements establish that evaluation of risks to health and safety and the measures needed to control them, will be matters for joint determination and negotiation. Through them, employers recognise the legitimacy of union-appointed health and safety representatives, together with the rights and powers they need to perform their roles effectively (the rights to inspect, to receive hazard information, to prevent a continuation of unsafe work). The agreements stipulate that all health and safety matters that would formerly have been raised by a worker with a supervisor will henceforth be channelled through the worker's health and safety representative, ensuring that such matters are dealt with on a collective rather than an individual basis. Furthermore, the agreements establish joint union–management health and safety committees at workplace level, company level and industry level as appropriate. Their constitution and terms of reference are spelt out in the agreements. Their functions are stipulated to be those of review, such as reviewing any disagreements that may have occurred between a health and safety representative and a supervisor, and of initiating the development of codes of practice and standards needed to enable work to proceed safely. Thus they function in the same manner as parliamentary committees, while the health and safety representatives are accountable to defined 'constituencies' of workers exposed to similar hazards in the same way that members of parliament are accountable to their constituents (Mathews, 1984; 1985b).

A key feature of the agreements is their provision for paid leave for training of health and safety representatives on union-nominated courses. This is in recognition of the fact that union representatives cannot be expected to negotiate on an equal footing with management unless they are given some grounding in hazard identification and control, and in the union and ACTU policies on this question. In this way the agreements establish a principle that can be incorporated in more comprehensive technological change agreements, providing for training of workplace representatives in key aspects of change and the unions' approach to it. Health and safety agreements are also important in another respect, in that they represent a consensus reached between unions and employers and thereby pave the way to legislation which extends the principles contained in the agreements to cover all workers.

I would argue that these agreements represent an important step in the direction of industrial democracy. The evaluation of proposed technological changes with regard to their likley hazards to health will evolve logically into wider evaluation of the effect of such changes on the labour process as a whole and, arguably from there, to an evaluation of their effects on the economy and society. An instance of such a wider evaluation occurred in the Victorian Railways in 1983. In the face of technological changes in the rail workshops that threatened extensive job losses and deskilling, the rail unions launched a sophisticated response within an overall consultative framework established by the new state Labor government. A union working group negotiated time off for some delegates who embarked on an extensive research and consultation exercise, publishing their proposals in a comprehensive report, *The future of the railway workshops* (ARU, 1983). The report's proposals gave the union side an initial position from which it was able to negotiate a sophisticated and constructive approach to questions of jobs, skills and demarcations in the workshops. This approach was then extended into the white-collar clerical area. When the issue of replacing the railways' antiquated telephone system came up, the union again had delegates booked off to prepare a case, which was published as *New job opportunities and training for V-Line's telephonists and teleprinter operators: an ARU response to technological change* (ARU, 1984). Again, the union was able to participate in discussions concerning a reorganisation of work on the basis of a prepared and considered position. The process was as important as the outcome.

The examples cited show that it is at least feasible, given a favourable political climate and sophisticated leadership, for unions to make progress in co-determining labour process issues in the workplace. This takes unions in a radically new direciton, away from the traditional oppositionist and abstentionist stances ('not doing the bosses' work'), and towards a constructive *but determined* approach to work reorganisation that involves workers themselves at every step, exercising collective pressure through their union. These changes in the workplace will not occur quickly or easily. There are ideologically opposed employers, aligned with the New Right, who will do all in their power (exercising economic, legal and political leverage) to uphold 'managerial prerogatives' at any price. There are managements who simply cannot see the consequences of involving workers in decisions, and are afraid of these issues; they will attempt to subvert the flexible potential of new technology in order to keep the hierarchies in the workplace intact. There are elements within the unions themselves who oppose co-determination as a form of collaboration (from the extreme left) or as giving excessive influence to the membership (from the extreme right). There are legal blocks to workers exercising influence, such as the existing companies legislation which requires boards of directors to act in the interests of shareholders but not necessarily with regard to the interests of workers. And there are the colossal problems posed by the imbalance in technical and economic resources between the parties involved in these new and technically demanding sorts of negotiations.

Nevertheless, in spite of these difficulties, there is a clear commitment from a number of unions to develop a framework of co-determination with employers on workplace issues, and to intervene in industry policy in a constructive manner. This can only take place within a favourable political climate such as was provided by the ACTU–ALP Accord, and policies emanating from it such as the federal government's Green Paper on *Industrial Democracy and Employee Participation* (1986). Underlying this political and ideological commitment, there are favourable developments in technology (for example, flexible manufacturing systems) that have the *potential* to provide workers with a wider and more responsible role — despite its being subverted in many cases (Shaiken et al., 1986). This potential is widely recognised as being a critical factor in industrial regeneration and recovery, as stated, for example, explicitly by the US economist Robert Reich in his influential text *The Next American Frontier*:

> The industries that will sustain the next stage of America's economic evolution will necessarily be based on a skilled, adaptable and innovative labour force and on a more flexible, less hierarchical organization of work. (Reich, 1983)

Ultimately, this shift to a new level of collective worker involvement in workplace issues, if it is not to become subverted in a new stage of more intensive exploitation, needs to be accompanied by a real and substantial shift to collective worker ownership of productive resources. There are indications that this long dream of the labour movement is finally coming about, not through State ownership and nationalisation, nor through individualised employee share ownership plans, but through Swedish-style employee investment funds (Meidner, 1976; Mathews, 1987). Workers who own the productive resources they work with, subject to due legal and social controls, are ultimately the best source for constructive technical innovations and productive work reorganisation. It is in opening up such a vista that the radical potential of a 'labour process' policy, implemented in an active way by trade unions, becomes manifest.

Conclusion

This chapter has argued that the current wave of industrial restructuring that is occurring in all advanced manufacturing countries, and is now starting to be felt acutely in Australia and New Zealand, need not follow an authoritarian, technologically determinist path. Public choices can be made. Technologies can be designed and installed in such a way that workers' skills are enhanced and their contribution to productivity is seen as central, or in such a way that human input is reduced to a minimum and the workers' role is seen as marginal. The labour process literature warns us not to fall into the trap of assuming that there is no way out for workers faced with authoritarian workplace technologies. This would be to foreclose the possibility of achieving further democratisation of social and economic life, which I take to be one of the basic aims of the labour movement this century. Rather, the new insights gained through the study of the labour process and managerial control techniques are now available to be applied to help workers and their unions counter the strategies deployed against them. This is the practical project which could only begin with the recent development of an adequate theorisation of the labour process.

Bibliography

Allen, S. and Wolkowitz, C. (1986) 'The Control of Women's Labour: The Case of Homeworking' *Feminist Review* No. 22, pp. 25–51

Anthony, P. D. (1977) *The Ideology of Work* London: Tavistock

Armstrong, P. (1986) 'Management control strategies and inter-professional competition: the cases of accountancy and personnel management' in D. Knights and H. Willmott (eds) *Managing the Labour Process* Aldershot: Gower

ATEA (1976) *Teletechnician* Melbourne: Australian Telecommunications Technicians Association

—— (1977) *Teletechnician* Melbourne: ATEA

—— (1982) *Teletechnician* Melbourne: ATEA

Atkinson, P., Reid M. and Sheldrake, P. (1977) 'Medical mystique' in *Sociology of Work and Occupations* Vol. 4, No. 3, pp. 243–80

Australian Bureau of Statistics (1974) *Retail Establishments and Selected Service Industries* Economic Census, 1968–1969

Australian Bureau of Statistics (1981) *Census of Retail Establishments and Selected Service Establishments* 1979–80

Australian Conciliation & Arbitration Commission (1984) *Decision: Termination, Change and Redundancy Case,* Print F6230, Melbourne, 2 August; *Supplementary Decision*, Print F7262, Sydney, 14 December

ACTU, Australian Council of Trade Unions (1987) *Future Strategies for the Trade Union Movement* Melbourne: ACTU D74–87

ARU, Australian Railways Union (1983) *The Future of the Railway Workshops* ARU: Melbourne

ARU, Australian Railways Union (1984) *New Job Opportunities and Training for V-Line's Telephonists and Teleprinter Operators: an ARU Response to Technological Change* ARU: Melbourne

ASTEC, Australian Science and Technology Council (1985) *Computer-related Technologies in the Metal Trades Industry* Canberra: Australian Government Printing Service

Badham, R. (1986) 'Technology and public choice: strategies for technological control and the selection of technologies' *Prometheus* Vol. 4 No. 2, pp. 288–305

Barker, J. and Downing, H. (1980) 'Word processing and the transformation of the patriarchal relations of control in the office' *Capital and Class* No. 10, pp. 64–99

Barr, T. (1985) *The Electronic Estate* Melbourne: Penguin

Baxter, V. (1979) 'Reorganisation of the Post Office in Historical Perspective' *Insurgent Sociologist* Vol. viii No. iv, pp. 17–32

Beechey, V. (1982) 'The sexual division of labour and the labour process a critical assessment of Braverman' in S. Wood (ed.) *The Degradation of Work?* London: Hutchinson

Bell, D. (1973) *The Coming of Post-Industrial Society: Adventure in social forecasting* New York: Basic Books

Benet, M. K. (1972) *Secretary: An Enquiry into the Female Ghetto* London: Sidgwick and Jackson

Bennett, E. C. (1979) *New Technology and the Australian Printing Industry* Sydney: PKIU

Berliner, H. (1985) *A System of Scientific Medicine* New York: Tavistock

Beynon, H. (1973) *Working for Ford* Harmondsworth: Penguin

Blackburn, P., Coombs, R. and Green, K. (1985) *Technology, Economic Growth and the Labour Process* London: Macmillan

Blauner, R. (1964) *Alienation and Freedom* New York: Wiley

Boreham, P., Clegg, S. and Dow, G. (1986) 'The institutional management of class politics: beyond the labour process and corporatist debates' in D. Knights and H. Willmott (eds) *Managing the Labour Process* Aldershot: Gower

Bradley, K., and Hill, S. (1983) '"After Japan": The quality circle transplant and productive efficiency' *British Journal of Industrial Relations* Vol. XXI, No. 3, pp. 291–311

Brailsford, J. F. (1952) 'Factors which influence the value of a radiological investigation ' *The Lancet* 5 April, p. 679

Braverman, H. (1974) *Labor and Monopoly Capital* New York: Monthly Review Press

Bray, M. and Taylor, V. (eds) (1986) *Managing Labour*? Sydney: McGraw Hill

Brecher, R. and Brecher, E. (1969) *The Rays: A History of Radiology in the United States and Canada* Baltimore: William Wilkins

British Medical Journal 1906, Special Correspondence, 27 January, p. 231

Brooks, R. (1983) 'The Melbourne Tailoresses' Strike, 1882–83: An Assessment' *Labour History* No. 44, May 1983, pp. 27–38

Brubaker, R. (1984) *The Limits of Rationality* London: Allen & Unwin

Bruland, T. (1982) 'Industrial conflict as a source of technical innovation: three cases' *Economy and Society* Vol. 11, pp. 91–121

Buchanan, D. A. (1986) 'Management objectives in technical change' in D. Knights and H. Willmott (eds) *Managing the Labour Process* Aldershot: Gower

Burawoy, M. (1979) *Manufacturing Consent: Changes in the Labor Process under Monopoly Capitalism* Chicago: University of Chicago Press

—— (1985) *The Politics of Production* London: Verson

Business Week (1984) 'The revival of productivity' 13 February, pp. 46–51

Butler, D. (1986) Secretarial Workers: A Study of their Changing Labour Process and Social Relations, unpublished PhD thesis, Brisbane: Griffith University

Cameron, R. (1937) 'The Victorian Boot and Shoe Industry' *Economic Record*, Vol. 13, pp. 31–5

Cammock, P. and Inkson, J. H. K. (1985) Technological change and industrial relations: a case study in the meat industry, unpublished paper, Department of Business Administration, University of Canterbury

Campbell (1983) Recent Marxist studies of the capitalist labour process: the limitations of the theme of capitalist control, unpublished MA thesis, La Trobe University

Child, J. (1985) 'Managerial strategies, new technology and the labour process' in D. Knights, H. Willmott and D. Collinson (eds) *Job Redesign* Aldershot: Gower, pp. 107–41

Clawson, D. (1980) *Bureaucracy and the Labour Process: The Transformation of US Industry 1860–1920* New York: Monthly Review Press

Clegg, S. and Dunkerley, D. (1980) *Organisation, Class and Control* London: Routledge and Kegan Paul

Cockburn, C. (1981) The material of male power *Feminist Review* 9, pp. 41–58; reprinted in, D. MacKenzie and J. Wajcman, (1985) *The Social Shaping of Technology* Milton Keynes: Open University Press

—— (1983) *Brothers: Male Dominance and Technological Change* London: Pluto Press

—— (1985) *Machinery of Dominance: Women Men and Technical Know-How* London: Pluto Press

Coghlan, T. A. (1918) *Labour and Industry in Australia* 4 vols, reprinted 1969 Oxford

Committee of Inquiry into Technological Change in Australia (CITCA) (1980) (Myers, Sir Robert et al.) *Technological Change in Australia* 4 vols, Canberra: Australian Government Publishing Service

Cooley, M. (1980) *Architect or Bee? The Human/Technology Relationship* Sydney: TransNational Co-operative Ltd.

—— (1984) *Technology, Unions and Human Needs* Geneva: International Metalworkers' Federation

Coombs, R. (1985) 'Automation, Managerial Strategies and Labour-Process Change' in D. Knights, H. Wilmott and D. Collinson (eds) *Job Redesign: Critical Perspectives on the Labour Process* Aldershot: Gower

Corrigan, D. (1828) 'Aneurism of the Aorta' *The Lancet* Vol. 1, pp. 586–90

Court Brown, W. M. and Doll, R. (1958) 'Expectations of life and mortality from cancer among British radiologists' *British Medical Journal* 26 July, pp. 181–7

Coventry, G., Cornish, G., Stricker, P., Cooke, R. and O'Brien, A. (1984) *Part-time Work and Youth in Transition* Melbourne: Victorian Institute of Secondary Education

Crozier, M. (1964) *The Bureaucratic Phenomenon* London: Tavistock

CSE Microelectronics Group (1980) *Microelectronics: Capitalist Technology and the Working Class* London: CSE Books

Daly, J. (1985) Assessing the hazards from diagnostic radiology unpublished M. Env. Sci. thesis, Melbourne: Monash University

Daly, J. and Willis, E. (1987) 'The social relations of medical technology: Implications for technology assessment and health policy' in J. Daly, K. Green and E. Willis (eds) *Health Care Technologies: Policies and Politics* Canberra: Australian Government Publishing Service

Daniel, A. (1986) 'Work in the Computer Industry' *Advance* Vol. 2, No. 2, pp. 9–10

Davidson, A. (1968) *Antonio Gramsci: The Man, His Ideas* Sydney: New Left Publications

Davison, G. (1978) *The Rise and Fall of Marvellous Melbourne* Melbourne: Melbourne University Press

Deery, S. (1982) 'Trade Unions, technological change and redundancy protection in Australia' *Journal of Industrial Relations* Vol. 24, No. 2, June, pp. 155–75

Department of Employment and Industrial Relations *Industrial Democracy and Employee Participation, Discussion Paper* Canberra: Australian Government Publishing Service

Dewar, D. L. (1980a) *The Quality Circle Handbook* Red Bluff: Quality Circle Institute

—— (1980b) *The Quality Circle: What You Should Know About It* Red Bluff: Quality Circle Institute

Dickson, D. (1974) *Alternative Technology and the Politics of Technical Change* Glasgow: Fontana

Dohse, K., Jurgens, V. and Malsch, T. (1985) 'From "Fordism" to "Toyotism"? The social organisation of the labor process in the Japanese automobile industry' *Politics and Society* Vol. 14, No. 2, pp. 115–46

Duckett, S. J. (1984) 'Structural interests and Australian health policy' *Social Science and Medicine* Vol. 18, No. 11, pp. 959–66

Dunford, R. W. (1983) 'Co-ordination or control: the split literature of organisational analyses', paper presented to Annual Conference of the Sociological Association of Australia and New Zealand, Melbourne, August

Dunkerley, D. (1980) 'Technological change and work: up-grading or deskilling' in P. Boreham and G. Dow (eds) *Work and Inequality* Vol. 2, Melbourne: Macmillan

Durkheim, E. (1957) *Professional Ethics and Civil Morals* London: Routledge and Kegan Paul

Economic and Budget Review Committee, Victorian Parliament (1986) Fourteenth Report: *A Labour Market Study for Radiologists*

Economic Research Unit Pty Ltd (ERU) (1980) 'Case studies on the impact of word processors, Vol. 1', Submission to Committee of Inquiry into Technological Change in Australia (CITCA) Melbourne

Eddy, C. E. (1946) 'Fifteenth anniversary of the discovery of x-rays' *Medical Journal of Australia* 2 February, pp. 138–41

Edsall, D. L. (1896) 'The attitude of the clinician in regard to exposing patients to the x-ray' *The Journal of the American Medical Association* Vol. 47, No. 18, pp. 1425–9

Edwards, P. (1986) *Conflict at Work* London: Blackwell

Edwards, R. (1979) *Contested Terrain: The Transformation of the Workplace in the Twentieth Century* New York: Basic Books

Eldridge, J. E. T. (1971) *Sociology and Industrial Life* London: Michael Joseph

Emery, F. E. and Phillips, C. (1976) *Living at Work* Canberra: Australian Government Publishing Service

Finch, J. and Groves, D. (1983) *A Labour of Love* London: Routledge and Kegan Paul

Fitzgerald, R. T. (1967) *The Printers of Melbourne* Melbourne: Sir Isaac Pitman and Sons Ltd.

Fortune Editors (1982) *Working Smarter* New York: Viking Press

Foulkes, F. K. and Hirsch J. L. (1984) 'People make robots work' *Harvard Business Review* Jan–Feb

Fox, A. (1971) *The Sociology of Work in Industry* London: Collier-Macmillan

Frances, R. (1986) '"No More Amazons": Gender and Work Process in the Victorian Clothing Industry, 1890–1939' *Labour History* No. 50, pp. 95–112

Frankel, B. (1983) *Beyond the State? Dominant Theories and Socialist Strategies* London: Macmillan

—— (1987) *The Post-Industrial Utopians* Cambridge: Polity Press

Freidson, E. (1970) *Professional Dominance* Chicago: Aldine

—— (1976) 'The division of labour as social interaction' *Social Problems* Vol. 23, pp. 304–13

Frenkel, S. J. (1987) 'Industrial sociology and workplace relations in advanced capitalist societies' *International Journal of Comparative Sociology* Vol. 27, Nos 1–2, pp. 69–86

Friedman, A. (1977a) *Industry and Labour* London: Macmillan

—— (1977b) 'Responsible autonomy versus direct control over the labour process' *Capital and Class* No. 1, Spring, pp. 43–57

Fry, E. (1956) 'Outwork in the Eighties' *University Studies in History and Politics* Vol. 2, July 1956, pp. 77–93

Game, A. and Pringle, R. (1983) *Gender at Work* Sydney: Allen & Unwin

Gill, C. (1985) *Work, Unemployment and the New Technology* Cambridge: Polity Press

Glenn, E. N. and Feldberg, R. L. (1979) 'Proletarianizing clerical work: technology and organizational control in the office' in A. Zimbalist (ed.) *Case Studies on the Labor Process* New York: Monthly Review Press

Goldthorpe, J., Lockwood, D., Bechhofer, F. and Plat, J. (1968) *The Affluent Worker: Industrial Attitudes and Behaviour* Cambridge: Cambridge University Press

Gramsci, A. (1971) *Prison Notebooks* New York: International Publishers

Green, D. and Cromwell, L. (1984) *Mutual Aid or Welfare State: Australia's Friendly Societies* Sydney: Allen & Unwin

Griffin, T. (1984) 'Technological change and craft control in the newspaper industry: an international comparison' *Cambridge Journal of Economics* Vol. 8, pp. 41–61

Grzyb, G. J. (1984) 'Tailored work groups: managerial recollectivisation and class conflict in the workplace' *Research in Social Movements, Conflict and Change* Vol. 6, pp. 143–66

Habermas, J. (1970) 'Technology and science as ideology' *Toward a Rational Society* Boston: Beacon Press

Hagan, J. (1966) *Printers and Politics* Canberra: ANU Press

—— (1983) 'Unions: context and perspective 1850–1980' in B. Ford and D. Plowman (eds) *Australian Unions* Melbourne: Macmillan

Hartley, J. (1983) *Robots at work: a practical guide for engineers and managers* Bedford (UK): IFS/North Holland

Hartmann, H. (1976) 'Capitalism, patriarchy, and job segregation by sex' in M. Blaxall and B. Reagan (eds) *Women and the Workplace* Chicago: University of Chicago Press

Hendricks, S. (1986) Industry Restructuring—A Case Study of the Australian Clothing Industry unpublished dissertation, Centre for Environmental and Urban Studies, Sydney: Macquarie University

Hernaman-Johnson, F. (1919) 'The place of the radiologist and his kindred in the world of medicine' *Archives of Radiology and Electrotherapy* Vol. 24, pp. 181–7

Hill, R. (1984) 'From hot metal to cold type: new technology in the newspaper industry' *New Zealand Journal of Industrial Relations* Vol. 9, No. 3, pp. 161–75

Hill, R. and Novitz, R. (1985) 'Class, gender and technological change: a working paper', paper presented to the Conference of the Sociological Association of Australia and New Zealand, University of Queensland, Brisbane

Hospital Radiation Technologists' Association of Victoria (1979) Submission and Draft Legislation on Radiographer Registration.

Hutchins, D. (1985) *Quality Circles Handbook* London: Pitman

Huws, U. (1984) 'The New Homeworkers—New Technology and the changing location of white collar work' *Low Pay Unit* Pamphlet No. 28, London

Hymes, D. (1974) *Foundations in Sociolinguistics: An Ethnographic Approach* Pennsylvania: University of Pennsylvania Press

Inkson, J. H. K. (1977a) 'The job satisfaction of New Zealand male manual workers' *New Zealand Psychologist* Vol. 6, No. 1, pp. 2–13

—— (1977b) 'The man on the dis-assembly line: New Zealand freezing workers' *Australian and New Zealand Journal of Sociology* Vol. 13, No. 1, pp. 2–11

—— (1978) 'A study of the aspirations of male manual workers in New Zealand': Occasional Paper No. 29, Department of Business Administration, Massey University

—— (1979a) 'Attitudes to industrial relations in four New Zealand industrial groups' *New Zealand Journal of Industrial Relations* Vol. 4, pp. 17–25

—— (1979b) 'Management practices and industrial conflict: the case of the

New Zealand meat industry' *New Zealand Journal of Business* Vol. 1, pp. 83–103

—— (1980) 'Factors influencing workers' involvement in their unions' *Journal of Industrial Relations* 22, pp. 442–52

—— (1987) 'The craft ideal and the integration of work: a study of potters' *Human Relations* Vol. 40, No. 3, pp. 163–76

Inkson, J. H. K. and Cammock, P. (1984) 'Labour process analysis and the chain system in the New Zealand meat works' *New Zealand Journal of Industrial Relations* Vol. 9, pp. 149–60

Inkson, J. H. K. and Simpson, D. (1975) 'The assembly-line and alienation: a participant–observer study in the meat industry' *New Zealand Psychologist* No. 4, pp. 44–55

ICRP, International Commission on Radiological Protection (1982) *Protection of the Patient in Diagnostic Radiology: Annals of the ICRP* Vol. 9, 2/3 (ICRP 34) Oxford: Pergamon

Jackson, E. et al. (1894) 'Report on the values objective tests for the determination of Ametropia, Ophthalmoscopy, Ophthalmometry, Skiascopy' *Journal of the American Medical Association* Vol. 88, pp. 337–9

Jalée, P. (1977) *How Capitalism Works* New York: Monthly Review Press

Jamous, J. and Peloille, P. (1970) 'Professions or self perpetuating systems: changes in the French hospital system' in J. Jackson, (ed.) *Professions and Professionalisation* Cambridge: Cambridge University Press

Jones, Bryn (1982) 'Destruction or redistribution of engineering skills? The case of numerical control' in S. Wood (ed.) *The Degradation of Work?* London: Hutchinson

Jones, Barry (1982) *Sleepers, Wake*! Melbourne: Oxford University Press

Jones, B. and Wood, S. (1985) Tacit skills, division of labour and new technology, unpublished manuscript

Journal of the American Medical Association 1903, 41, pp. 499

Juravich, T. (1985) *Chaos on the Shop Floor: A Worker's View of Quality, Productivity and Management* Philadelphia: Temple University Press

Kahn, R. L. (1972) 'The meaning of work: interpretation and proposals for measurement' in A. Campbell and P. E. Converse (eds) *The Human Meaning of Social Change* New York: Russell Sage Foundation

Katz, H. C. and Sabel, C. F. (1985) 'Industrial relations and industrial agreement in the car industry' *Industrial Relations* Vol. 24, No. 3

Kelly, J. (1982) *Scientific Management, Job Redesign and Work Performance* New York: Academic Press

—— (1985) 'Management's redesign of work: labour process, labour markets and product markets' in D. Knights, H. Willmott and D. Collinson (eds) *Job Redesign* Aldershot: Gower

Kirby, P. (Chairperson) (1985) *Report of the Committee of Inquiry into Labour Market Programs* Canberra: Australian Government Publishing Service

Knights, D. and Collinson, D. (1985) 'Redesigning work on the shopfloor: a question of control or consent?' in D. Knights, H. Willmott and D. Collinson (eds) *Job Redesign* Aldershot: Gower

Knights, D. and Willmott, H. (eds) (1975) *Gender and the labour process* Aldershot: Gower

Knights, D. and Willmott, H. (eds) (1986) *Managing the Labour Process* Aldershot: Gower

Knights, D., Willmott, H. and Collinson, D. (eds) (1985) *Job Redesign: Critical Perspectives on the Labour Process* Aldershot: Gower

Komhauser, A. (1965) *Mental Health of the Industrial Worker* Chicago: Wiley

Kraft, P. and Dubnoff, S. (1986) 'Job Content, ·Fragmentation, and Control in Computer Software Work' *Industrial Relations* Vol. 25, No. 2, pp. 184–96

Kreigler, R. (1979) *Working for the Company: work and control in the Whyalla Shipyard* Melbourne: Oxford University Press

Kumar, K. (1977) *Prophecy and Progress* Harmondsworth: Penguin

Labour Research (1986) 'Fast Food—faster profits' No. 75, pp. 10–13

Larkin, G. (1983) *Occupational Monopoly and Modern Medicine* London: Tavistock

Larsen, M. (1977) *The Rise of Professionalism: A Sociological Analysis* Berkeley: University of California Press

Laslett, P. (1971) *The World We Have Lost* New York: Scribners

Lawler, E. E. and Mohrman, S. E. (1985) 'Quality circles after the fad' *Harvard Business Review* Jan–Feb.

Lawson, T. (1981) 'Paternalism and labour market segmentation theory' in F. Wilkinson (ed.) *The Dynamics of Labour Market Segmentation* New York: Academic Press

Lazonick, W. (1979) 'Industrial relations and technical change: the case of the self-acting mule' *Cambridge Journal of Economics* Vol. 96, pp. 89–109

—— (1981) 'Production relations, labor productivity and choice of technique: British and US cotton spinning' *The Journal of Economic History* Vol. XLI, No. 3, pp. 491–516

Lee, D. (1982) 'Beyond Deskilling: skill, craft and class' in S. Wood (ed.) *The Degradation of Work?* London: Hutchinson

Linge, G. J. R. (1979) *Industrial Awakening: A Geography of Australian Manufacturing, 1788–1890* Canberra: ANU Press

Lipsig-Mummé, C. (1983) 'The Renaissance of Homeworking in Developed Economies' *Relations Industrielles* Vol. 38, No. 3, pp. 545–66

Littler, C. (1982) *The Development of the Labour Process in Capitalist Societies: a Comparative Study of the Transformation of Work Organization in Britain, Japan and the USA* London: Heinemann Educational

Littler, C. (1987) 'What difference has the labour process debate made?' paper presented to the Conference of the Sociological Association of Australia and New Zealand, Sydney

Littler, C. R. and Salaman, G. (1982) 'Braverman and beyond: recent theories of the labour process' *Sociology* Vol. 16, No. 2, pp. 251–69

Lloyd, C. (1985) *Profession: Journalist* Sydney: Hale and Iremonger

Locke, E. A. (1976) 'The nature and causes of job satisfaction' in M.

Dunette (ed.) *Handbook of Industrial and Organizational Psychology* Chicago: Rand McNally, pp. 1297–1349

Macdonald, R. (1982) *David Syme* Melbourne: Vantage House

MacKenzie, D. and Wajcman, J. (eds) (1985) *The Social Shaping of Technology: How the refrigerator got its Hum* Milton Keynes: Open University Press

Macklin, J. (1982) *Part-time Work: Not for the State Bank* Melbourne: Labor Resource Centre

Manwaring, T. (1982) 'The trade union response to new technology' *Industrial Relations Journal* Vol. 13, No. 1, pp. 7–26

Manwaring, T. and Wood, S. (1985) 'The ghost in the labour process' in D. Knights, H. Willmott and D. Collinson (eds) *Job Redesign* Aldershot: Gower

Marsden, D. et al. (1985) *The Car Industry: Labour Relations and Industrial Adjustment* London: Tavistock

Marshall, A. (1983) *Changing the Word* London: Comedia Publishing Group

Martin, R. (1981) *New Technology and Industrial Relations in Fleet Street* Oxford: Clarendon Press

—— (1984) 'New technology and industrial relations in Fleet Street: "new technology will make it possible for managers to manage"' in M. Warner (ed.) *Microprocessors, Manpower and Society* Aldershot: Gower

Marx, K. (1967) *Capital* Moscow: Progress Publishers

Mathews, J. (Chairperson) (1984) *The Protection of Workers' Health and Safety: Report of the Steering Committee* (3 volumes) Adelaide: South Australian Department of Labour

—— (1985a) *Technology, Trade Unions and the Labour Process* Working papers in the social studies of science, No. 1, Department of Humanities, Deakin University, Vic; reproduced in Submission to Senate Standing Committee on Science, Technology and the Environment, Inquiry into new technology and employment *Hansard* Senate Hearings: Melbourne, 28 August, pp. 1632–1813

—— (1985b) *Health and Safety at Work: Australian Trade Union Safety Representatives' Handbook* Sydney: Pluto Press Australia

—— (1986) *The Transformation of Work: Technological Change and Union Strategies of Intervention* Working papers in the social studies of science, No. 2, Department of Humanities, Deakin University, Victoria

—— (1987) 'The push for collective investment' *Australian Society* Vol. 6, No. 1, pp. 19–21

McClenahan, J. L. (1970) 'Wasted x-rays' *Radiology* Vol. 96, August, pp. 453–6

McGraw, P., Dunford, R. and Knights, S. (1987) *Quality Circles, Employee Participation and Industrial Relations: A Study of Australian Industry* Employee Participation Report No. 11, Canberra: Australian Government Publishing Service

McNally, F. (1979) *Women for Hire: A Study of the Female Office Worker* London: Macmillan Press

Meidner, R. (1976) *Employee Investment Funds: an Approach to Collective*

Capital Formation London: Allen & Unwin (English translation 1978)
Melman, S. (1983) *Profits Without Production* New York: Alfred A. Knopf
Merrington, J. (1968) 'Theory and practice in Gramsci's Marxism' in R. Miliband, and J. Saville, (eds) *Socialist Register* London: Monthly Review Press
Murphy, R. (1986) 'Weberian closure theory: a contribution to the ongoing assessment' *The British Journal of Sociology* Vol. 37, No. 1, pp. 21–4
Noble, D. (1979) 'Social choice in machine design: the case of automatically controlled machine tools' in A. Zimbalist (ed.) *Case Studies on the Labour Process* New York: Monthly Review Press
—— (1984) *Forces of Production: a Social History of Industrial Automation* New York: Alfred A. Knopf
O'Donnell, C. (1984) *The Basis of the Bargain* Sydney: Allen & Unwin
OECD (1986) *Employment Outlook*, September, pp. 15–65
Olson, M. and Primps, S. (1984) 'Working at Home with Computers' *Journal of Social Issues* Vol. 40, No. 3, pp. 97–112
Palmer, L. S. J. (1985) Technology and Telecommunications: Labour Process and Forms of Intercourse in the Transformation of Australian Telecommunications, unpublished PhD thesis, La Trobe University
—— (1986a) 'Telecom, Information Technology and the Labour Process,' APROS Conference, Melbourne University, May
—— (1986b) 'Extensions to Labour Process Theory', SAANZ Conference, University of New England, July
Parkin, F. (1979) *Marxism and Class Theory: A Bourgeois Critique* London: Tavistock Publications
Penn, R. D. (1985) Socialization into skilled identities: an analysis of a neglected phenomenon, unpublished paper, Department of Sociology, Lancaster University
Pratt, A. (1908) *David Syme, The Father of Protection in Australia* Melbourne: Ward Lock and Co. Ltd
Pratt, J. (1984) 'Home Teleworking' *Technology Forecasting and Social Change* Vol. 25, pp. 1–14
Radford, E. P. (1981) 'Cancer risks from ionizing radiation' *Technology Review* November/December, pp. 74–5
Ramsay, H. (1985) 'What is Participation For? A critical evaluation of the labour process analyses of job reform' in D. Knights, H. Wilmott and D. Collinson (eds) *Job Redesign: Critical Perspectives on the Labour Process* Aldershot: Gower
Ray, C. A. (1986) 'Corporate culture: the last frontier of control' *Journal of Management Studies* Vol. 23, No. 3, pp. 287–97
Reed, M. I. (1984) 'Management as a social practice' *Journal of Management Studies* Vol. 21, No. 3, pp. 273–85
Reed, R. (1987) 'Making newspapers pay: employment of women's skills in newspaper production' *The Journal of Industrial Relations* Vol. 29, No. 1, pp. 25–40
Reeders, E. (1986) 'Students as Workers: the Effects of Big Mac' *Youth Studies Bulletin* Vol. 5, No. 1, pp. 31–5

Reich, R. (1983) *The Next American Frontier* New York: Times Books

Reinecke, I. (1982) *Micro Invaders* Ringwood: Penguin Books

—— (1983) *The Phone Book* Melbourne: Penguin

—— (1985) *Connecting You* Melbourne: Penguin

Reiser, S. J. (1978) *Medicine and the Reign of Technology* Cambridge: Cambridge University Press

Richardson, J. (1987) 'Technology assessment in medicine: an Australian proposal' in J. Daly, K. Green and E. Willis (eds) *Technologies in Health Care: Policies and Politics* Canberra: Australian Government Publishing Service

Riches, L. (1980) 'The dimensions of technological change and trade union responses' *The Australian Quarterly* Vol. 52, pp. 191–212

Rieter, E. (1987) 'The Labour Process in the Fast Food Industry—A Feminist Perspective on Work' Aston—UMIST Labour Process Conference, Manchester

Rimmer, P. J. (1974) 'The Boot and Shoe Industry in Melbourne' in J. M. Powell (ed.) *Urban and Industrial Australia* Melbourne: Sorrett

Roberts, M. (1980) 'Sally Bowen: Political and Social Experience of a working class woman' in E. Windschuttle (ed.) *Women, Class and History* Sydney: Fontana/Collins

Robson, M. (ed.) (1984) *Quality Circles in Action* Aldershot: Gower

Robson, P. (1980) *An Assessment of Work Organization Associated with ARE 11 Communications Technology* Melbourne: Australian Telecommunications Employees' Association

Rose, M. and Jones, B. (1985) 'Managerial strategy and trade union responses in work reorganisation schemes at establishment level' in D. Knights, H. Willmott and D. Collinson (eds) *Job Redesign* Aldershot: Gower

RACR, Royal Australian College of Radiologists, 1987, May mini budget changes Newsletter No. 2

Rubery, J. (1980) 'Structured labour markets, worker organization and low pay' in A. Amsden (ed.) *The Economics of Women and Work* London: Penguin, pp. 242–70

Ryan, E. and Prendergast, H. (1982) 'Unions are for women too' in K. Cole (ed.) *Power, Conflict and Control in Australian Trade Unions* Ringwood: Penguin

Sabel, C. (1982) *Work and Politics* Cambridge: Cambridge University Press

Salaman, G. (1979) *Work Organisations, Resistance and Control* London: Longman

—— (1982) 'Managing the Frontier of Control' in A. Giddens and D. MacKenzie (eds) *Social Class and the Division of Labour* London: Cambridge University Press

Sayers, C. E. (1965) *David Syme, A Life* Melbourne: Cheshire

Scarbrough, H. (1984) 'Maintenance workers and new technology: the case of Longbridge' *Industrial Relations Journal* Vol. 15, No. 4, pp. 9–17

Schudson, M. (1980) 'Review of Larson's "Rise of Professionalism"', *Theory and Society* Vol. 9, pp. 215–19

Senate Standing Committee on Science, Technology and the Environment (1987) *Technology Assessment in Australia* Canberra: Australian Government Publishing Service
Serle, G. (1971) *The Rush to be Rich* Melbourne: Melbourne University Press
Sethi, S. P., Namiki, N. and Swanson, C. L. (1984) *The False Promise of the Japanese Miracle* Boston: Pitman
Shaiken, H. (1984) *Work Transformed: Automation and Labor in the Computer Age* New York: Holt, Rinehart and Winston
Shaiken, H. Herzenberg, S. and Kuhn, S. (1986) 'The work process under more flexible production' *Industrial Relations* Vol. 25, No. 2, pp. 167–83
Shanks, S. C. (1950) 'Fifty years of radiology' *British Medical Journal* 7 January, pp. 44–8
Skinner, W. (1986) 'The productivity paradox' *Harvard Business Review* July–August
Smith, A. (1976) *The Wealth of Nations* London: Oxford University Press
—— (1980) *Goodbye Gutenberg* New York: Oxford University Press
Souter, G. (1981) *Company of Heralds* Melbourne: Melbourne University Press
SPRU Women and Technology Studies Group (1982) *Micro-electronics and Women's Employment in Britain* SPRU Occasional Paper Series 17, Sussex
Stark, D. (1980) 'Class struggle and the transformation of the labour process' *Theory and Society* Vol. 9, No. 1, pp. 89–130
Storey, J. (1985a) 'The means of management control' *Sociology* Vol. 19, No. 2, pp. 193–211
—— (1985b) 'Management control as a bridging concept' *Journal of Management Studies* Vol. 22, No. 3, pp. 269–91
—— (1986) 'The phoney war? new office technology: organisation and control' in D. Knights and H. Willmott (eds) *Managing the Labour Process* Aldershot: Gower
Sutcliffe, J. T. (1921) *A History of Trade Unionism in Australia* (reprinted 1967) Sydney: Macmillan
Svedberg, B. (1986) 'Technology and human resources', keynote address to second World Conference on *Productivity and the Future of Work* Munich, October 1986
Sweet, R. (1983) 'Changing Patterns of Work and Education' in D. S. Anderson and C. Blakers (eds) *Youth, Transition, and Social Research* Canberra: ANU Press
Sweezy, P. (1974) 'Introduction' in H. Braverman *Labour and Monopoly Capital* New York: Monthly Review Press
Taylor, F. W. (1947) *Scientific Management* New York: Harper
Telecom (1982) *Memorandum of Understandings* Melbourne: Telecom Australia Central Office
Thompson, E. P. (1968) *The Making of the English Working Class* Harmondsworth, Penguin Books
Thompson, P. (1983) *The Nature of Work: An Introduction to debates on the labour process* London: Macmillan

Thompson, P. C. (1982) *Quality Circles* New York: AMACOM
Toffler, A. (1980) *The Third Wave* London: Collins/Pan
Tolliday, S. and Zeitlin, J. (eds) (1986) *The Automobile Industry and its Workers: Between Fordism and Flexibility* Cambridge: Polity Press
Turkington, D. J. (1976) *Industrial Conflict: a Study of Three New Zealand Industries* Wellington: Methuen
Turner, B. A. (1984) 'Sociological aspects of organisational symbolism', paper presented to the Department of Sociology, University of Trento, Trento, Italy, November
Vedel, G. (1985) 'Telematics and Remote Office Work' Institute for Industrial Research and Social Develpment, Denmark
Versluysen, M. (1981) 'Midwives, medical men and "poor women labouring of child" Lying in hospitals in eighteenth century London', in J. Roberts, (ed.) *Women, Health and Reproduction* London: Routledge and Kegan Paul
Victorian Government: Report to the Premier of Victoria (1981) *Report of the Inquiry into the Ownership and Control of Newspapers in Victoria* by the Hon. J. G. Norris (The Norris Report)
Voydanoff, P. (1978) 'The relationship between perceived job characteristics and job satisfaction among occupational status groups' *Sociology of Work and Occupations* Vol. 5, No. 2, pp. 179–92
Wajcman, J. (1983) *Women in Control* Milton Keynes: Open University Press
Walker, C. R. and Guest, R. H. (1952) *Man on the Assembly Line* Cambridge (Mass.): Harvard University Press
Weber, M. (1947) *The Theory of Social and Economic Organization* translated by A. M. Henderson and T. Parsons, New York: The Free Press
Wilkinson, B. (1983) *The Shopfloor Politics of New Technology* London: Heinemann
Williams, C. (1981) *Open Cut* Sydney: Allen & Unwin
Willis, E. (1983) *Medical Dominance: The Division of Labour in Australian Health Care* Sydney: Allen & Unwin
—— (1985) 'Trade union reaction to technological change: the introduction of the chain system of slaughtering in the meat export industry' *Prometheus* Vol. 3, No. 1, pp. 51–70
—— (1986) 'RSI as a social process' *Community Health Studies* Vol. 10, No. 2, pp. 210–19
Wilson, B., Wyn, J., Reeders, E. and Woock, R. (1986) *Education, Work and Youth Policy* Canberra: Commonwealth Schools Commission
Windschuttle, K. (1986) 'Education, High Technology and the Economy' in R. Castle, D. Lewis and J. Mangan (eds) *Work, Leisure and Technology* Melbourne: Longman Cheshire
Winner, L. (1985) 'Do artifacts have politics?' in D. MacKenzie and J. Wajcman (eds) *The Social Shaping of Technology* Milton Keynes: Open University Press
Wochos, J. F., Detorie, N. A. and Cameron, J. R. (1979) 'Patient exposure from diagnostic x-rays—an analysis of 1972–1975 NEXT data' *Health Physics* Vol. 36, pp. 127–34

Wood, S. (ed.) (1982) *The Degradation of Work?: Skill, deskilling and the labour process* London: Hutchinson

Wright Mills, C. (1951) *White Collar* Oxford: Oxford University Press

Zimbalist, A. (1979) 'Technology and the labor process in the printing industry' in A. Zimbalist (ed.) *Case Studies on the Labor Process* New York: Monthly Review Press

—— (1979) *Case Studies on the Labor Process* New York: Monthly Review Press

Index